The Leader's Playbook
5 Building Blocks for Creating a Quality-First Culture

Chad Dorgan

QUALITY PRESS

Milwaukee, Wisconsin

American Society for Quality, Quality Press, Milwaukee 53203
© 2024 by ASQ
All rights reserved. Published 2024
Printed in the United States of America

28 27 26 25 24 LS 5 4 3 2 1

Publisher's Cataloging-in-Publication data

Names: Dorgan, Chad B., 1966-, author.

Title: The leader's playbook : 5 building blocks for creating a quality-first culture / Chad Dorgan.

Description: Includes bibliographical references. | Milwaukee, WI: Quality Press, 2024.

Identifiers: LCCN: 2024949735 | ISBN: 9781636941851 (paperback) | 9781636941868 (PDF) | 9781636941875 (epub)

Subjects: LCSH Leadership. | Corporate culture. | Organizational effectiveness. | BISAC BUSINESS & ECONOMICS / Leadership | BUSINESS & ECONOMICS / Workplace Culture

Classification: LCC HD30.28 .D67 2024 | DDC 658.4—dc23

ASQ advances individual, organizational, and community excellence worldwide through learning, quality improvement, and knowledge exchange.

Bookstores, wholesalers, schools, libraries, businesses, and organizations: Quality Press books are available at quantity discounts for bulk purchases for business, trade, or educational uses. For more information, please contact Quality Press at 800-248-1946 or ask@asq.org.

To place orders or browse the selection of all Quality Press titles, visit our website at: http://www.asq.org/quality-press

Printed in the United States of America.

Quality Press
600 N. Plankinton Ave.
Milwaukee, WI 53203-2914
Email: books@asq.org
Excellence Through Quality

Contents

Preface .... *v*
Acknowledgments .... *ix*
Introduction .... *xi*

Part I: Foundational and Performance Elements **1**

Chapter 1 Foundational Elements → Performance Element 3
Chapter 2 Driving Value 17
Chapter 3 Foundational Element 1: Right ("RT1") 27
Chapter 4 Foundational Element 2: Individual/Collaboration 45
Chapter 5 Foundational Element 3: Embedded Verification 59
Chapter 6 Foundational Element 4: Continuous Improvement 71
Chapter 7 Performance Element: Everyone Benefits 91

Part II: Catalysts **97**

Chapter 8   Catalyst 1: The Power of Once 99
Chapter 9   Catalyst 2: Data Intelligence 107
Chapter 10 Catalyst 3: Knowledge Snippets 117

Part III: Your Quality First Journey **125**

Chapter 11 A Road Map to Quality First 127

Conclusion **153**

Appendices: Quality First Perspectives **157**

Appendix A   Construction Quality First Example.   159

Appendix B   Manufacturing Quality First Example   163

References *165*

Preface

My journey to creating this book, like most things, started out very small. As a young lieutenant in the U.S. Air Force, I was responsible for designing a retrofit of the heating and cooling systems for the hospital's operating rooms at the base where I was stationed. Then, because I was deployed to Operation Desert Storm, I missed the entire construction of the project. However, after I returned from deployment there were several middle-of-the-night phone calls from the base's general that the system was not working. Fortunately, we were able to get the systems working.

My thoughts at the time centered around the following question: How can a great design with great contractors end up with something that doesn't work, especially with all the quality checks that are integrated into the process of design and construction on an Air Force project? Several decades later, it is easy to see that having quality-focused processes, procedures, and activities often does not mean you will achieve high quality on a consistent basis or for the long term.

I have also, unfortunately, witnessed and been directly affected by organizations that were clearly recognized as leaders in quality for their industries, but due to changes in them and their leadership, their quality departments were eliminated or downgraded to the point where the organizations themselves no longer drove from a quality perspective. These organizations went from the top in their industries for quality to near the bottom in a very short time, and then often took a decade or more to return to the top.

In addition, during my years in the field, I have read, followed, and implemented the key lessons from dozens of business and quality books. They all have great points, structures, and ideas, and they made me a much better manager and leader. However, there was always a gap between the business side and the quality side—primarily because, for many leaders, quality is a department or element of the organization, but it rarely comes up when discussing how to truly run and drive the organization. Leaders will often adopt quality approaches and tools, such as strategic quality planning, root cause analysis, Hoshin planning, or any number of others, but they will not address quality from the core of their business.

This is where *The Leader's Playbook* and the principle of Quality First comes in. To be clear, Quality First is not another quality process to add to your business. Quality First, in this book, is defined as "establishing quality at the very center of the organization and having it flow from and through the organization's core—its values and mission—so that quality provides a consistent foundation for all departments, individuals, initiatives, and projects in the organization to determine focus, execute their roles at the highest level, and continually improve how they operate."

There are many different things a company could consider to be "first." Maybe your company is in technology, and innovation is essential to you, or you are a manufacturer, and materials and supply chain are first. However, as you consider becoming and leading a Quality First organization, it is important to understand that Quality First must become the foundation of other initiatives and focus areas to be successful. Quality First will improve your strategic focus and tactical implementation by ensuring quality is truly integrated into the *what* and *how* of your organization—it becomes the natural way you operate instead of layers added to address issues caused when quality is not at your core. A good way to think about this is that in Quality First organizations, quality becomes a proactive and positive reality to which you are naturally drawn, versus the current state of many organizations where quality is seen as something very reactive—trying to find and resolve issues—and thus becomes something people want to avoid.

Let's look at an example of how having Quality First as your foundation applies to implementing new initiatives. One that is near and dear to me is sustainability. Say your company is just embarking on its sustainability and corporate social responsibility journey. By starting with and using Quality First, you already have clarity on a common language and approach to aligning your core values and mission to operations and continuous improvement. Bringing sustainability to this quality foundation simplifies and accelerates its successful integration into all your organization's operations, departments, and projects. Conversely, when not using Quality First, sustainability initiatives often focus on gathering data and creating reports and publications to get people excited but not on making real change. Thus, with Quality First as your sustainability foundation, you will have an immediate and long-term impact on your organization, community, and the environment.

A final key element to remember about Quality First is that it is not necessarily a destination you reach but a journey of continually refining and improving as your organization grows and changes. Quality First provides you with the terminology and approach to place quality at the core of your organization, so it flows through every part of the enterprise. However, the specifics of how you define and ingrain Quality First into your organization will be unique to you, just as your organization is unique from all others.

I am excited that you are about to begin leading your organization to put Quality First. Like everything in life, it is not hard, but it also is not easy. As you read this book, please take time to consider how the key points and information in the text boxes and activities apply to you and your organization. If you can, form a small group to help you do this; a group provides the best and quickest way to gain insights and drive the needed change.

Chad Dorgan
Yorba Linda, California

Acknowledgments

So many individuals and organizations influenced me over the years and provided elements that ultimately became *The Leader's Playbook* and Quality First. All interactions provided diverse and essential insights, and I thank everyone who helped on this journey.

My mom and dad, Joan and Chuck Dorgan, have always been there to support and push me, and they were the ones who knew I could always do more. I can't tell you how my mom's congratulatory words after I spoke at a conference or training gave me the desire to do it again, even though I thought I did just OK—I miss her dearly. My dad has been my boss, mentor, and role model—from challenging me in high school with complex energy audits as part of his company to showing me that volunteering and giving back are just as important as the work you do—he continues to guide and support me.

My first leaders and bosses in the Air Force provided me with the ability to try new ideas and push the envelope on what was acceptable at the time; their faith in me, and giving me room to stumble and learn, gave me a great appreciation of leadership and organizations. Thank you to Colonel Don Meister, Captain Bob Gaias, and John Sabochick.

As the young leader of Dorgan Associates, I think I learned more from my employees than they could ever have learned from me. It is with them and through them that Quality First really began taking shape. These include the original three—Joy Altwies, Ian MacIntosh, and Svein Morner—along with Steve Leight, and, of course, the glue that held us all together, Mary Sue Quigley. Also, a special shout-out to Michael Savone, Sidney Parsons, and especially Chad Grindle—all men with hearts of gold who left us too soon.

I also want to thank Mike Hurst and Derek Glanvill for bringing me in as their new quality leader, where I was able to put Quality First to the test and hone it to what it is today. Their guidance, patience, and willingness to think outside the box allowed me to spread my wings and grow as a leader.

I also owe incredible gratitude to Sue Klawans and Scott West, my original compadres in creating the Construction Quality Executives Council, which is now the Design and Construction Excellence Exchange (DCX) Quality Exchange. Their support, guidance, critiques, and extremely long conversations over dinner and drinks solidified Quality First concepts. A special remembrance for Scott West—he left this world way too early, and his legacy will live on through us all. I would also like to recognize the founding members for supporting this journey: Mike Clippinger, Bill Vandrovec, Peter Ukstins, Mike Stadjuhar, and Terry Brinkman, for their undying support of the organization and our journey, and Rodney Spencley, for being the true contrarian and keeping us honest.

Finally, I would like to thank my wife Christine and my children. Nicholas and Rachel. You give me all the love and support I will ever need. Being your husband and father fills me with joy, and I hope *The Leader's Playbook* makes the world a better place for us to cherish and enjoy with family and friends.

Contributors' Acknowledgments

The American Society for Quality (ASQ) and Quality Press would like to thank the Quality Press Peer Review Committee for its invaluable volunteer participation and contributions to this work. Without our volunteers' subject matter expertise, time, and passion for creating content, none of our efforts would be possible.

2023–2026 Committee Members

- Scott A. Laman, Chair
- Jane Keathley, Vice Chair
- Melvin Alexander
- Ahmad Elshennawy
- Marc Hamilton
- Gary Jing
- Trevor Jordan
- Peter Pylipow
- G.S. Sureshchandar
- Tiea Theurer
- Mary McShane-Vaughn

Introduction

Through my personal Quality First journey, I have realized that quality cannot be driven by an individual or department. Quality is not a process, procedure, or checklist—these are only tools for achieving quality. In addition, especially for an organization's quality department, quality cannot only be focused on one primary aspect of the business, such as manufacturing or operations. For quality to truly be integral and add the immense value it potentially can to an organization, it must come from and be part of the organization's DNA.

Think about your own organization or other organizations that you admire—what drives them and what defines them? Inherently, organizations are defined and driven by their core values—those elements that are developed over time based on the actions and norms of individuals. Core values are not created by a group of people sitting around a table; rather, they come from the behaviors of the organization as a whole.

By starting with your organization's core values, which may need to be updated to reflect the reality of your organization (but more on that later in the book), and aligning and integrating quality, your entire organization can begin to plan, operate, and behave in a Quality First manner from and through your core values. Let's think about this a little before moving on. If your core values are just how you act, then starting with them and aligning quality at the core provides the framework whereby quality just becomes how you act.

This is the elegance and simplicity of Quality First—when it comes from your organization's core, it is not an additional layer or major change you need to add. You don't have to "become lean," earn your Black Belt, or even become an expert on control charts, fishbone diagrams, or Pareto charts. Though all these approaches and tools are useful and can help you improve, they do not address the fundamental issue: they are added onto an organization instead of being the organization. Many organizations out there have adopted lean, Six Sigma, and other approaches that have improved their functioning. However, because these approaches did not come from the core of the company, many of these organizations have also slid back due to a change of leadership, market downturns, and other occurrences that degraded or eliminated the initiatives. (See Appendix B.)

To best enable you to start, lead, and go through your Quality First journey, follow this primer (Figure 0.1). It consists of four foundational elements that form the basis of Quality First; one performance element, which is a result of implementing the foundational elements; and three catalysts that supercharge your Quality First journey, as well as the creation of a road map, the use of key tools, and the presence of engaged leadership.

This book is divided into three parts and appendices:

- *Part I: Foundational and Performance Elements.* At the heart of Quality First are four foundational elements—right, individual/collaboration, embedded verification, and continuous improvement—that form the basis of approaching and implementing Quality First from your core, along with one performance element (everyone benefits), which is the result of implementing the four foundational elements. A thorough understanding and internalization of these elements, both personally and within your organization, is required before you can start your journey.

- *Part II: Catalysts.* Three catalysts for Quality First are essential for the foundational elements to thrive in your organization. The first catalyst is "the power of once," which embeds the drive to do all work once throughout your organization. The second catalyst is "data intelligence," which uses data, data analytics, and business intelligence to utilize data and visualization to identify high-risk/high-opportunity areas for the foundational elements. The third catalyst is "knowledge snippets," which allows the right information to get to the right individuals at the right time so they can do their work right the first time—once.

- *Part III: Your Quality First Journey.* A road map always makes it easier to get to your destination, so this section provides you with a Quality First road map and timelines that have been proven to work and are easily adaptable to your organization.

- *Appendices: Quality First Examples.* The two appendices to the book contain Quality First examples and the value brought to organizations and industries.

QUALITY FIRST PRIMER

What is Quality First?
Quality First supercharges your organization from your core values through integration and consistent use of the Foundational Elements

FOCUS ON 4 - 1 - 3

Leading Your Organization to be Quality First requires focusing on the 4 - 1 - 3 components.

4 Fully integrate the 4 Foundational Elements into your organization starting from and through your core values:

1. Right
2. Individual/Collaboration
3. Imbedded Verification
4. Continuous Improvement

1 Highlight and communicate the Performance Element of Everyone Benefits to demonstrate and drive success.

3 Utilize the 3 Catalysts to accelerate your Quality First Journey:

1. Power of Once
2. Data Intelligence
3. Knowledge Snippets

ROAD MAP
Create a simple road map to take you on your Quality First Journey, focusing on 1 - 5 - 10 year milestones.

KEY TOOLS
- Agile Teams and Collaborative Planning
- 1-Page Communication
- Nominal Group Technique
- Business Intelligence

LEADERSHIP
Fully engage leadership throughout the organization, from C-Suite/Board to Department Heads and Champions.

Figure 0.1 Quality First primer.

Please remember that Quality First is a journey that comes from your organization's core and must be an integral part of any changes to that core, including strategic planning initiatives, mergers and acquisitions, and significant growth or reduction periods. Quality First not only improves each of these but also ultimately provides the initial framework and discussion for the constant change your organization undergoes.

Finally, as you embark on your Quality First journey, please make it *your* journey. Although I provide guidance, support, and insight along the way, as you will find from the second foundational element, we always focus on the individual. It is important that the terms used, the way they are implemented, and the way you continuously improve make sense to you and your organization. For example, if RT1 is not a term that resonates with your organization (see discussion in Chapter 3), use one that conveys the same concept but works for you.

Let your Quality First journey begin!

Part I

Foundational and Performance Elements

Chapter 1
Foundational Elements → Performance Element

If you are new to quality, and even for some of us who have 30 or more years working in quality, the terminology used and the varying number of approaches available are confusing, to say the least, if not downright dizzying. Given that most quality approaches and programs have been developed either for a specific purpose (such as the Toyota Production System—lean), for a specific industry (for example, Six Sigma—manufacturing), or even to fill a perceived void or need (for example, commissioning process—owner facilities), they have each developed their own catch-phrases and terminologies.

However, by simply stepping back from each of the quality approaches out there, it is possible to see clear alignment and consistencies that permeate all quality approaches—I call these *elements*, which are fundamental to this book's concepts. Though I admit this is another new term, it is probably the most important term and concept to learn, given that it truly forms the base for your organization to be the best it can be and supercharges everything you do.

There are two types of elements: foundational elements, which form the basis for all quality programs and are the focus of your Quality First journey, and a performance element, which is the result of integrating the foundational elements into your organization. These elements comprise a five-point framework, as shown in Figure 1.1.

- Foundational elements:
    1. Right ("RT1")
    2. Individual/collaboration
    3. Embedded verification
    4. Continuous improvement

- Performance element:
    1. Everyone benefits

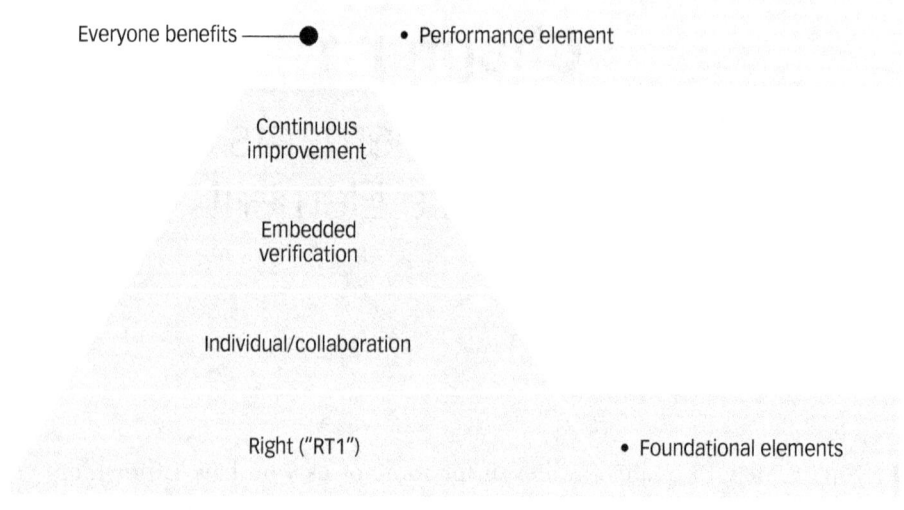

Figure 1.1 The five elements.

Let's look at each foundational element and the performance element.

Foundational Element 1: Right

The cornerstone of my Quality First concept is a fundamental element I reference throughout the book as "right." For the sake of clarity and to not confuse the word *right* with the foundational element, I will denote this element as "RT1" from this point forward. An easier way to understand and think about the first foundational element is to expand RT1 to be:

"How do we define and do RT1?"

In this context, *RT1* is about how to achieve an anticipated outcome or result as correct. This simple question is actually very challenging to answer and to keep in the forefront of everything you do in your organization. Yet it is also elegant, in that RT1 is an innate focus for all of us—we want to do our work right, whether the work direction is clear or not. There is a reason why RT1 is the first foundational element from both a quality and a cultural perspective, which is that doing work correctly transcends the quality process and management programs that have already been tried. While it may have been called something else, such as waste elimination, defect-free, reliability, zero incidents, or others, ultimately quality is striving to always do the work right once.

The reason RT1 is the first foundational element is that it focuses on knowing what the result is going to be across all aspects of the business and then working on how to get there, which is the "do right" part. Often, when looking at major failures within an organization, the root cause of the issue was not that the process was broken or that the people were trying to do something wrong, but that they never took the time to define what RT1 was and then use that as their guiding light.

By integrating RT1 (define and do) across all parts of the business, you establish clarity on, and an expectation of, empowering individuals to achieve RT1 in their work. This is the true power and beauty of RT1, in that the specific definition of what RT1 is varies from the organizational level to the departmental level, all the way down to the individual level. Just as each individual is unique, so is his or her understanding and focus on what is "right." That uniqueness, collectively, is what truly supercharges the organization, and there can be a hierarchy of RT1, with the lower-level RT1s often aligning with those at the organizational level. By being complementary to one another, it helps drive the organization's purpose and values more successfully (see Figure 1.2).

Organization
1. Achieve long-term positive results for our stakeholders.
2. Be a responsible member of the community we work in.
3. Continuously reduce our impact on the environment for future generations.

Human resources
- Maintain parity in all positions for all employees (Organization 1).
- Align workforce to that of the community where we work (Organization 2).

Sustainability
- Ensure the business does not harm the environment but rather regenerates it (Organization 3).
- Promote and welcome others to follow our lead (Organization 2).

Manufacturing
- Eliminate, redesign, or manage by-products of manufacturing to minimize environmental impact (Organization 3 and Sustainability 1).
- Promote continuous improvement on achieving best-in-class products (Organization 1).
- Work with local schools to provide internships for manufacturing skills (Organization 2).

Financial
- Maintain growth in top and bottom lines (Organization 1).
- Provide transparency of financials to employees (Human Resources 1).
- Performance rewards based on improving environment (Sustainability 1).

Figure 1.2 Example of an RT1 hierarchy.

It is important for RT1 to entail making money and being successful—if a business does not do these things, it cannot continue to operate. However, if you define RT1 as only making money, you miss the opportunity to be the best version you could be—to be exceptional. By expanding and integrating discussions of RT1 throughout your organization, you begin to form the foundation for success—you define RT1 and then enable your people to do RT1.

To achieve this, it is important to verbalize RT1 and discuss it openly—it cannot be assumed everyone knows and is aligned with RT1. By integrating RT1 into the way you strategize, plan, and run your organization, you are taking the first step to Quality First and to supercharging your organization.

Foundational Element 2: Individual/Collaboration

Think about going to dinner at your favorite restaurant. You have just ordered your favorite meal and are looking forward to a great evening. As you relax and enjoy yourself, the meals around you are being served, and they look amazing. Then your meal is served, and it looks amazing, too. You cut into it, take your first bite, and instantly spit it out—the inside is ice cold.

Now, aside from ruining your night and likely getting you pretty mad, let's take a look at this ice-cold meal from a quality perspective. In preparing the ingredients and your meal, four or more individuals were likely involved. I'm an eternal optimist and truly believe that no one purposely tries to mess up and do wrong (see the first foundational element), so at the individual level, all individuals want to do their job correctly.

But when making a meal, what is RT1? Obviously, it is not making a frozen meal, but to "achieve" RT1, the kitchen staff may have had to:

- Transfer the food from the freezer to the refrigerator two days ago
- Remove the food from the refrigerator in the morning and start marinating it
- Season the food and put it in a pan to sear
- Put the pan in an oven to cook to temperature
- Verify the temperature before serving

At each of these steps, an individual has to do his or her work right for the final product, your meal, to be RT1. In addition, along the way, a simple change or misstep results in issues—what if the thermometer is broken or misreads the temperature?

The crucial point is that individuals ultimately determine the level of quality achieved—either they get it right or they don't. Anyone else who is involved after their point of work can only identify issues or mistakes, not determine the quality achieved, for that was done by the individual. This point cannot be overemphasized because, unfortunately, many companies focus their efforts only on verification and inspection. Inspection finds the issues *after* they have been made, resulting in rework, not in avoiding the occurrence.

Individual/collaboration is a foundational element because once you understand and internalize that it is the individuals who determine the quality, your focus changes from a reactive one (finding mistakes) to an enabling one (how do I know that):

- The individual knows what RT1 is for his/her work.
- The individual has the information and knowledge to do RT1.
- The individual has the right tools and materials at the right time to do RT1.
- The individual can verify and communicate he/she did RT1 to the next individual.

However, individuals by themselves are islands. This is where collaboration, the other part of the second foundational element, comes in. Defining RT1 and making sure the appropriate information, tools, and materials are available to the individual at the correct time requires collaboration by many people. This is called many different things, such as planning, procurement, and training—but ultimately, it is collaboration around the individual that achieves success (Figure 1.3).

A good way to think about collaboration is to picture a dartboard, with the individual being the bullseye. There need to be many activities and people around the bullseye for that individual to be successful. All these parts come together when needed for the individual to do his/her job and coordinate with others. In addition, the individual is responsible, at a minimum, for coordinating and collaborating with the individuals upstream and the individuals downstream from his/her activity.

Manufacturing. I think the most classic example of this interaction is in manufacturing, where there is a linear series of activities being accomplished.

- *Individual:* If we look at an individual on the line, he/she has to be trained to know what to do (what is RT1), to understand what the correct starting point is (what he/she is receiving), and what the correct end point is (what to deliver to the next person). Ultimately, even with this knowledge, it is still the individual who determines the quality that is delivered—it is either RT1, or it is not. This is why understanding and focusing on the individual is so important.
- *Collaboration:* I like using the manufacturing line as an example because you also quickly see that individuals must collaborate and rely on those around them to be able to do their job right. This could be interacting with different people, using visual cues to communicate, or having daily stand-up meetings as a team to discuss how they continually improve.

Building construction. A trade professional on a construction site—say, installing carpet—demonstrates where the work is not as linear and is often not repeatable, as compared to a manufacturing line.

- *Individual:* At the individual level, one must understand what RT1 is again, but that changes for every project and could vary from room to room with different materials or patterns. However, it is again up to the individual to actually do the work right—if it is not aligned correctly, terminated at the edges, and so on, it will be wrong.
- *Collaboration:* There is actually more than you think—the trades that pour the concrete floor have to use the right mix, finish it properly, and cure it properly. The floor must be flat and level and have the correct moisture content for the flooring to adhere to it. In addition, there is also coordination downstream for others to protect the flooring from damage from subsequent work so the carpeting remains RT1.

Figure 1.3 Examples of individual–collaboration interaction.

Foundational Element 3: Embedded Verification

Verification and inspection are at the heart of most quality programs, as they enable checking work against known standards or goals. However, a common challenge is that the focus and timing of these activities often results in finding issues after they have occurred, resulting in rework and increased waste. Think about the restaurant example, where the food was ice cold. In this situation, you were the person doing the verification after all the work was done—and you found that a major mistake had been made. The only way to fix this is to start the entire process over again.

Unfortunately, this is the approach many companies take—they use verification and inspection to find mistakes that have already been made, not to avoid them in the first place. This is where the third foundational element comes in: *embedded verification*.

The difference with embedded verification is simply that it is accomplished by those doing the work when they do the work, relative to the definition of RT1 for the work. In a traditional sense, embedded verification uses and complements existing quality tools, such as visual controls, measurements, and poka-yoke, but it is intended to extend into all parts of the company—from human resources to procurement.

To understand embedded verification better, let's take a look at how the first two foundational elements relate to embedded verification.

- RT1: As should be clear by now, the first step in understanding RT1 is defining it. Therefore, once it has been defined, you can verify against it. Depending on the criticality of an item, this verification could range from a couple of questions to an extremely detailed checklist (for example, a preflight checklist for a pilot).

- Individual/Collaboration: Because the individual is the one who determines the quality achieved, that person needs to know what RT1 is and document that he/she has achieved it. It is also important to communicate to those upstream if issues are found (the previous individual may need to change) and to those downstream (to show the work was done right).

Therefore, *embedded verification* occurs as a result of the individual's effort and is clearly communicated with the larger group (collaboration) (see Figure 1.4). Ideally, embedded verification is accomplished in a three-step (Triple A) process: alignment, accomplish, and acknowledge.

1. **A**lignment on RT1: Before doing the work, the individual (and team) align around what RT1 is and how it will be documented. For repeated activities, this is often documented in a checklist or procedure to which the individual is trained and oriented. For one-time activities that are not repeatable, the process may be less formal, but it still needs to be discussed, and criteria should be documented.

2. **A**ccomplish Work: Upon understanding RT1 and how the individual will document the work, the individual completes the task. The concept is similar to taking a practice test in school—you understand what you will be graded on, so you will perform better. When you don't understand what success is, you often don't achieve it.

3. **A**cknowledge RT1: After the work is completed, the individual who did the work reviews the criteria and documents that it was done correctly. Again, depending on the criticality of the activity, the level of documentation could vary from a simple electronic checkbox to a signature on a card to requiring precise measurements and documentation.

Human resources
- Key item list to check before sending offer letter for new employee.
- Call checklist to verify issue was resolved (or what to do if not).

Financial
- Expense report automatically verifies attached receipts.
- Individual answers two or three questions before submitting his/her expense report.

Barista
- Repeat order after written on cup.
- Repeat order when delivered to customer.

Figure 1.4 Examples of embedded verification.

The focus, goal, and intent of embedded verification are to ensure these are just things the individual does as part of his/her job that become habits, so no one passes on work that is not RT1. While the hope is that you can eliminate, or at least minimize, the need for third-party/postproduction verification or inspections, in reality, you want a constant balance. Ultimately, I have found that the more you drive embedded verification to the individual, the less onerous the traditional inspections become. However, whole industries have been built on finding problems. If there are no problems, they are still incentivized to find them. Be aware of this as you journey to Quality First and make sure you instead adopt embedded verification.

Foundational Element 4: Continuous Improvement

Every quality program is grounded in *continuous improvement*. This is called many different things with varying levels of robustness and structure to them, but fundamentally: if you don't improve how you do something every day, it is not a quality process. Continuous improvement is not easy, because it is never-ending. The vast majority of businesses and management professionals prefer to approach improvement in a stepwise fashion. They focus on a process, improve it, and then wait for a period of time, often a substantial one, before addressing it again. The problem with stepwise improvement is that even when you improve it, there is always an opportunity to improve it more.

Put another way, nothing is ever perfect. One industry that epitomizes continuous improvement is computer chip manufacturing. At their core, these manufacturers understand that they follow Moore's Law; if they do not continually push themselves and find a better way to design and manufacture their chips, someone else will do so, and they will not be in business long (see Figure 1.5).

Moore's Law is an empirical relationship based on the fact that every two years, the number of transistors doubled on an integrated circuit board. The same trend has been found in computer storage, bandwidth speed, and solar cell capacity.

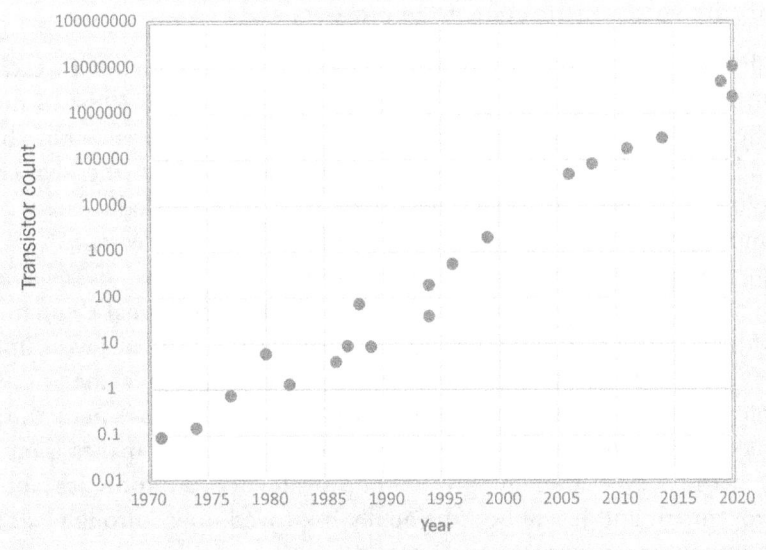

Figure 1.5   Moore's Law.

Moore's Law has been shown to work in almost any industry, in that the natural improvement of something has a factor of two every two years—which could be that something becomes twice as fast or half as expensive.

Therefore, if we can truly achieve continuous improvement in everything we do, it should generally follow Moore's Law of doubling (or halving) every two years. Now, I know you are thinking: My business is not making computer chips and cannot in any way follow this same trend!

My answer is: It can, but you need to approach improvement continuously—something that every person and every team discusses and does every day. Implementing what is highlighted by Paul Akers in his book *2 Second Lean: How to Grow People and Build a Fun Lean Culture at Work & Home* (Akers 2011) is a great option. While there is a lot of detail in his book, from its title you can tell that its focus is on improving what you do every day by two seconds.

The application of continuous improvement is pretty easy—the challenging part is to do it every day with focus and purpose. Failures to continuously improve typically result from a lack of consistency (daily) or follow-through (seeing long-term results). Therefore, continuous improvement in Quality First means to accomplish quick daily meetings/stand-ups that:

- Review what you were going to improve yesterday—did it work or not, and what was learned?
- Identify what you are going to improve today.

These daily meetings/stand-ups should be done at the team level, with each individual participating, and the entire meeting taking no more than 10 or 15 minutes. Periodically, it is important to pause and discuss continuous improvement in a broader context and identify the value to the individual and the team—whether that's money saved, avoided time spent, waste elimination, or other. These periodic pauses provide ongoing motivation for the daily meetings and continued improvement.

A quick note: Continuous improvement is accomplished by everyone in the organization, which goes back to the individual foundational element—it is the individuals who determine quality, and they are also the ones who improve. Therefore, the only way to improve continuously is to make the individuals responsible. This does not supplant the larger process improvement initiatives and formal continuous improvement frameworks; it actually makes them more powerful, because even after the process improvement, you know it's not just a step improvement, and it will be continually improved upon through execution, due to your continuous improvement culture.

Performance Element: Everyone Benefits

While looking at the foundational elements, you should begin to see an interconnection between them and understand that they all need to work together to be successful. If there is degradation in one of the foundational elements, they all suffer. Figure 1.6 provides an example of how they are interconnected.

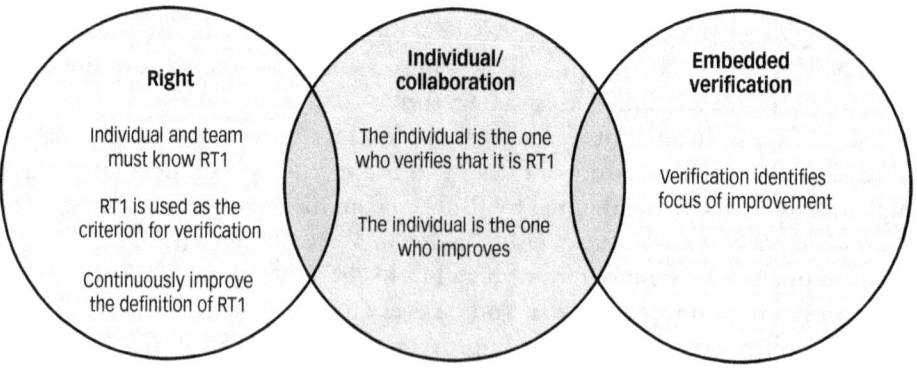

Figure 1.6 Interconnection of the foundational elements.

When you fully integrate the foundational elements into the way you operate, the results will be substantial and very positive, with the four foundational elements achieving the performance element of *everyone benefits*. However, it is important to explicitly focus on making sure everyone benefits to achieve the true potential of Quality First. With this type of focus, you ensure that Quality First remains broad and includes all stakeholders—internal to internal, and internal to external.

Given that the value of making sure everyone benefits comes from the foundational elements, here are some simple examples of benefits achieved through the different foundational elements:

- RT1: Clearly defining and communicating RT1 throughout the organization and projects helps maintain focus on the key outcomes and aids in the transparency of expectations and results. By keeping the commonly understood definition of RT1, misunderstandings are reduced and discussions shift to achievement versus blocking.
- Individual/collaboration: Individuals determine quality, yet they need the collaboration of others to streamline planning, simplify hand-off points, and empower the individual to execute RT1.
- Embedded verification: This is probably where the greatest value is gained, because the individual knows what is the right thing to do and verifies RT1 for his/her work. This eliminates the potential for rework and mistakes—which are typically between 20 and 30 percent in even the best processes. When issues arise, embedded verification allows for quick remediation and adaptation to avoid future issues.
- Continuous improvement: Truly continuous improvement is a game changer for any organization and empowers individuals to improve what they do every day. Their work becomes easier and more enjoyable, which leads to a better quality of life.

Ensuring that everyone benefits requires some level of focus and attention from leadership—it cannot be assumed or ignored. When Quality First is fully ingrained in your organization and everything is working, there is a tendency for leaders and others to become complacent and forget why it is working so well. This has, unfortunately, happened in too many organizations to count; an established quality culture can change, almost overnight, due to a significant change in leadership. In most cases, the new leader does not understand the need to do something (daily stand-ups, for example) and just eliminates it. Put another way, the organization has many of the Quality First elements, but it has not become Quality First from its core.

It's sometimes hard to see the broader value of Quality First in a single organization. As part of creating and refining Quality First over the past decade, I was fortunate to work with quality peers in the construction industry, where we adopted and internalized the foundational elements into our companies. The results are dramatic when comparing those following Quality First to other companies in the industry (see Figure 1.7).

As I said previously, once you compromise one element, you compromise them all, and you are no longer Quality First. Therefore, the value of focusing on ensuring that everyone benefits is that it provides the data, information, and storytelling within the organization that help protect, maintain, and drive the foundational elements. Although this may seem a bit self-serving, remember that past organizations that lost their focus went from being leaders in their industries to dropping to the bottom of the pack in about a year. It then takes about a decade to recover from such an experience and get back to the top (see the manufacturing example in Appendix B).

Figure 1.7 A snapshot of Quality First construction organizations.

There are multiple ways to document how everyone benefits, and many of these are likely already common practice in your organization. The key is to ensure that someone is consolidating the results and providing a formal summary each year to senior leadership and all individuals. Some opportunities to document this include:

- Surveys. Simply asking for feedback from your employees, customers, suppliers, and other stakeholders provides you with an understanding of what is working and what is not working. Specific questions about each foundational element can gauge your journey on the Quality First path.
- Industry rankings. Using industry rankings is another great metric (see Figure 1.7). This can range from revenue to the best places to work. Typically, the more third-party the evaluation, the greater the value.
- Employee performance plans. Integrating a discussion into each employee's performance review is a great way to get immediate feedback at the individual level. At a minimum, this should include a discussion of how the employee continuously improved throughout the year and the value he/she and the organization received from the improvement.
- Business intelligence. Using business intelligence on your internal systems to identify, visualize, and track trends relevant to the foundational elements can be beneficial.

Good or bad, there is not one way to identify and communicate how everyone benefits in your organization. The keys are to have a program to address it and to discuss it at all levels in the organization, including the boardroom.

Chapter 2
Driving Value

Before taking a closer look at each foundational element and how to integrate it into your organization, it is important to understand the value proposition of achieving Quality First in your organization. "Value" is composed of a definition portion and a driving portion—both of which need to work together to produce value. The power of Quality First is that the foundational elements set the groundwork for having the proper discussions required at all levels of the organization to engage in a common language for value, purpose, and direction. However, the actual definition and driving of value will vary based on your business segment and purpose.

Defining Value

In any organization, both direct and indirect types of value are produced. While most organizations understand and focus on direct value—such as goods produced, stock price, or employee turnover—it is just as important, if not more important, to include indirect value, which is often much harder to measure. This could include stakeholder satisfaction, employee wellness, or product life-cycle impact.

Obviously, to drive value, it is best to first define and understand what the value is. Starting with the foundational elements provides the common language and reduces duplicative efforts. The best way is to start with how the organization or a segment of the organization has documented what is RT1 for them relative to the first foundational element, because this typically encompasses a good deal of value statements within it.

To define value from an RT1 statement, you can utilize a 5-W approach, which is similar to the root cause tool of 5 Whys, used to understand why something did not work—but with a twist. The 5-W approach asks five questions based on "W" relative to value in order to frame the value; the questions allow the organization or segment to get clarity on the values that can be driven through achieving RT1. The 5-Ws are:

1. **W**hat happens when RT1 is achieved?
2. **W**hy is this important?
3. **W**ho benefits?
4. **W**hen is the value generated?
5. **W**ays to improve the value?

The order in which the W questions are answered does not matter; they just need to be discussed and answered. I recommend that the 5-W discussion be incorporated into the creation of your RT1 statements, because it will enable a deeper understanding and commitment by those involved, as well as provide an improved way to communicate to others the value of RT1.

W-1: What Happens When RT1 Is Achieved?

Typically, when RT1 is achieved, there are multiple benefits to the organization and the broader stakeholder group, including direct and indirect benefits. There may also be some potential real or perceived negative items, such as the decrease in need for personnel due to improved efficiencies. These should also be documented and discussed to provide understanding and an opportunity to address issues throughout the process, not after the impact has occurred.

W-2: Why Is This Important?

The second question to dive into is why it is important to achieve RT1; or, put another way, what happens if RT1 is not achieved? This may or may not be the opposite of what was determined during the W-1 stage, as the consequences could be much larger, such as going out of business or losing significant market share.

I recommend that you ask "Why" a couple of times to "peel the onion back" a bit and break through some of the unfounded assumptions you may hold. For example, if you state that it is important because "RT1 provides an opportunity to a marginalized community," then the next question would be "Why is that important?" This is the classic 5 Whys approach, embedded in your value discussion.

W-3: Who Benefits?

Understanding who benefits, or who is potentially affected, is important as we address the broader stakeholder group within the value discussion. When identifying who, it is best to look at these categories:

- Employees
- Shareholders
- Customers
- Supply chain
- Local community
- Broader community

W-4: When Is the Value Generated?

This W question will probably be one of the harder ones to answer, because it is often very fluid when true value is generated. While making a part and selling it provides "initial value" to the organization, it is not the total value because you are not accounting for returns, warranty issues, or even end-of-life disposal by the person who bought the product.

It is important to include a timeline approach to the "when" discussion, where you cover the life cycle of RT1 and value, at least as a starting point. There will be many discussions that only reveal immediate value. However, you'll be surprised by how often you discover many hidden values in the timeline that no one had ever considered.

The benefit of this timeline approach is that you can understand and potentially reap greater value by extending the timeline of your focus. This is especially true as organizations move into a circular economy approach, with some organizations now leasing their products so they retain the raw material value within their control. This shifts the value proposition from selling the product to selling, recovering, and reusing the materials in the product; while this could be a 10-year or longer cycle, the value propositions are often easily 3 to 10 times of just selling the product.

W-5: Ways to Improve the Value?

The final W question asks you to discuss how to improve the value achieved. This is intended to be the bridge between defining and driving value, as it begins the continuous improvement loop on achieving and improving upon RT1 and the subsequent value.

Given that the definition of value is so varied, examples of ways to improve value will be just as varied. A few questions to consider discussing are:
- What are reasonable one-year and five-year value targets?
- Is there a limit to the value that can be achieved?
- How can we accelerate value?
- What are the limiters to achieving value?
- Which values are complementary?

Driving Value

Once RT1 and value have been defined and understood, we then shift to driving value. This is where the second foundational element comes in—focus on the individual and collaboration to drive and achieve RT1 and value.

Ultimately, value is generated and determined at the individual level, and just like the broader focus on quality, collaboration between individuals is required to maximize that value. Therefore, to drive value through the individual, we need to address communicating, tracking, continually improving, and providing recognition of the value produced.

Communicating

Communicating could be called gaining alignment, understanding, or even commitment from the individual on what RT1 is and what the value of achieving it is. As RT1 and value go hand in hand, including value in the discussion is one of the best ways to gain commitment from individuals to do right, as they begin to see and understand that there is direct value to them.

The biggest challenge in communicating RT1 and value is that it is hard to do, and for the majority of what we do, we overcomplicate the message. It often becomes more about value to others versus value to the individual. When creating messaging on value, it's important to remain in the first person and be able to answer the traditional "What is in it for me?"

The answer does not need to be complicated or grandiose, but it does need to be personal whenever possible. A best practice that I have found to be highly effective is to focus on the conversation versus on me presenting a bulleted list of items that I think constitute the value. The conversation could encompass:

- Let's look at XXX RT1—what does this mean to you?
- If you do not achieve XXX RT1, how does this directly affect you?
- If you do not achieve XXX RT1, how does this affect others?
- OK, what is the value of you achieving XXX RT1?

Throughout this discussion, you will find what concerns the individual, but you will also get these individuals to directly define their personal value of doing RT1. Therefore, the key to communication is that it is an ongoing process, done by individuals with other individuals. The fact that the conversations are happening is more important than the details of the conversations.

Tracking

The old adage that if you don't measure it, it won't improve is true when driving value. The challenge with tracking value is that the effort to track it cannot outweigh the value gained.

Avoiding metric and tracking overload requires discipline and limiting what is tracked. A good guideline to follow is to limit metrics to no more than three per business segment, which should roll up to no more than one or two for each organizational level of RT1. This is challenging, as leaders love metrics and numbers to run their organizations; however, I find that the more metrics you have, the less effective they become. By limiting the number of metrics you observe, you can find a deeper understanding of the data and the metrics, thus providing a greater impact on operations and improvement.

Additional guidance on tracking is given in Chapters 8-10.

Continually Improving

Once you have a metric and can track it, you can continually improve your performance vis-à-vis the metric. To be successful, this needs to be reasonable and achievable. By reasonable, I mean that the rate of improvement over a specified period is something your team members can look at and easily say "Yes, we can do that." To be achievable, this is the end goal of improvement over the period; again, individuals can see that over a one-, three-, or five-year period, they can reach the goal.

Visually showing the metrics and trends is recommended, and it should include any degradation in improving value. Of course, it is important to address any degradation and not have it continue, but it is OK and natural to have setbacks in anything we do.

As is covered in greater detail in Chapter 4 on the individual/collaboration foundational element, having daily, small, continuous improvements will be the best way to implement tracking.

Recognition

The last item to drive value is to recognize that value has been gained. This can be formal recognition through celebrations, performance plans, bonuses, and promotions. It can also be informal through a thank-you or email message or as a discussion item during meetings.

What is important is that recognition is given. Without recognition, individuals and groups will think, "Why should I do anything when no one notices?" and they will begin to act on that thought. Recognition is the ongoing focus on RT1 and value, and recognizing the improvement will keep your teams engaged and continually improving.

Value Examples

Every part of an organization improves how employees plan and execute their work when they become Quality First, which includes the C-suite/executive level that focuses on and discusses the identification and driving of value throughout the organization and understands that how it flows through the organization will vary. For illustration purposes, we'll look at examples in areas such as operations, finance, human resources, social responsibility, safety, and information technology to provide a better understanding and context to defining and driving value.

Operations

In looking at operations for an organization, many times RT1 is focused on how productions are improved and waste is minimized. For this example, let's let RT1 be defined generically as "continuous improvement in achieving best-in-class products." The 5-Ws:

1. What: We maintain leading-edge products that have the highest level of quality, while always improving the processes to make the products.

2. Why: Stagnation kills a product and an organization—by combining the focus on continuous improvement and best in class, balance is achieved between the needs of the consumer and the ability to produce.

3. Who: The consumer benefits from higher-quality products that meet their needs, and the organization benefits through streamlining of processes and continual improvement of production by eliminating waste and rework.

4. When: The value is created throughout the entire life cycle, including minimization of waste, higher quality with lower recall/warranty rates, and improved consumer satisfaction.

5. Ways: Value can be improved by aligning and including the voice of each key stakeholder, including the consumer, designer, supplier, and manufacturer.

Value of RT1 Statement: By continuously improving our best-in-class products, all stakeholders benefit through alignment on purpose, reduction of production costs through streamlining of processes and eliminating waste and rework, and minimization of recalls and warranty requests.

Finances

For the financial health of an organization, RT1 can be described as "maintaining growth in the top and bottom lines." The 5-Ws:

1. What: When the top line of the organization is improved, more revenue is generated by the organization; when the bottom line of the organization is improved, the effort (overhead) required to create profit is decreased.
2. Why: It is important to focus on both top and bottom lines because there is a synergy between them—if the bottom line is too high, there is no capital to invest in the organization and its people, yet the top line needs to be high enough to support both the bottom line and growth.
3. Who: The organization benefits by having greater financial flexibility, the shareholders benefit with better returns on investment, and the employees benefit with greater benefits and bonuses.
4. When: Value is created in both the short and long terms.
5. Ways: Value can be improved through a continual focus on balancing the top and bottom lines based on current conditions and future trends.

Value of RT1: Through a balance and focus on the top line of the organization and the efforts required to generate that (overhead), there is improved value for the organization, shareholders, and employees by having sufficient equity for growth and investing in the organization's people.

Human Resources

There are many ways to define RT1 for human resources. Consider defining RT1 as "maintaining parity in all positions for all employees." The 5-Ws:

1. What: With parity, opportunity, pay, and benefits are equal for everyone at every level.
2. Why: This is important because there will be no actual or perceived gaps between individuals based on anything except objective performance.
3. Who: The organization and individuals benefit because the organization will attract the best and brightest, and individuals will benefit knowing that they are treated equally and fairly.

4. When: The value is generated every day by eliminating tension and disproportion between individuals and groups.

5. Ways: Value can be improved through greater transparency and open discussion.

Value of RT1: Parity between groups and individuals at all levels of the organization provides for equal opportunity, benefits, and pay, and removes actual and perceived barriers and gaps, making for an inclusive and vibrant community.

Social Responsibility

Social responsibility can span from the environment to interacting with the community around you. For this example, let's look at RT1 being "our business does not harm the environment; it regenerates it." The 5-Ws:

1. What: Every aspect of the business contributes to regenerating the environment.

2. Why: With limited natural resources and the continual degradation of the environment worldwide, actually regenerating the environment helps to turn the tide.

3. Who: Everyone benefits, yet by driving regeneration and circularity, organizations have been found to be more profitable and resilient compared with others.

4. When: Value can be immediate, but most is long term, especially on a community and global scale.

5. Ways: Addressing and continually changing processes to become more regenerative.

Value of RT1: By transforming the organization to be regenerative, we will directly contribute to turning the tide on the long-term damage done by humankind on the environment and will make our organization more profitable and resilient to future challenges.

Safety

Safety is more than just preventing people from getting hurt. RT1 for safety can be "we improve the wellness of our employees and visitors." The 5-Ws:

1. What: By improving the wellness of everyone, work becomes someplace you want to be because you are improving yourself, not just avoiding being hurt.

2. Why: There are many stressors on our workers and visitors—by addressing these and minimizing or eliminating them, the wellness of individuals is improved, making them more productive and valuable to the organization.

3. Who: Anyone who interacts with the organization can be affected, and those who are more directly linked will have greater improvement.

4. When: The value is both short and long term, with short term being from a single visit and long term being changed lifestyles and behaviors.

5. Ways: Continually evaluating products, systems, and food consumed in the organization to improve the wellness of both employees and visitors.

Value of RT1: By focusing on improving the wellness of our employees and visitors, we produce an environment they want to be in, which results in greater productivity and resources for the organization and long-term individual benefits to our employees.

Information Technology

When looking at RT1 in relation to the information technology of an organization, a good focus can be that "technology enhances the work experience." The 5-Ws:

1. What: By focusing RT1 on enhancing the work experience, it extends the discussion to actually using the technology.

2. Why: Technology only brings value when it is fully utilized and avoids adopting technology for technology's sake.

3. Who: The individual user becomes more productive and is enabled to maximize the value of the technology investment.

4. When: There are immediate and long-term values.

5. Ways: Incorporate continuous feedback loops on the technology, including the use of key performance indicators on the technology's success.

Value of RT1: Technology is right for the organization when it enhances the work experience of all individuals so they understand its use and purpose and can be fully productive.

In the next chapter, we'll dive into the first foundational element—RT1.

Chapter 3
Foundational Element 1: Right ("RT1")

As we begin to take an in-depth look into each foundational element, it is important to remember the context of the five elements and that RT1 is the first foundational element, as it sets the groundwork and definition of success for the remaining four. It is also one of the hardest to tackle, as many people feel they know what RT1 is and will do it, even if the definition is unclear.

As you begin your journey using the RT1 foundational element, remember that it is a journey and not a destination. What is defined as RT1 can always be improved upon and will always be changing throughout a project, as well as over time. Being flexible and adaptable to the journey is what will make it successful for you, your teams, and your organization.

Importance to Quality First

To make your organization Quality First, you must start with a discussion and the definition of RT1 for your organization because this sets or reaffirms the direction in which you are going and provides a common vernacular for everyone to use. As Quality First comes from and is core to your organization, having clarity and alignment on what RT1 is enables the start of your Quality First transformation.

Therefore, as you work your way through this chapter, take notes on how you will define and do RT1 throughout your organization and begin to engage others in the discussion as your first step to Quality First.

Defining RT1

Defining RT1 needs to become a standard practice and discussion within your organization at all levels. To be successful, RT1 must flow from your organization's core values, given that those values already define RT1 for your organization. However, unlike your core values, which typically stay at the organizational level, RT1 must be adapted and translated for each department, segment, and project to make it meaningful and addressable by the individuals doing the work.

To be clear, RT1 at the department level flows from and aligns with RT1 at the organizational level—it just includes more succinct and clear statements of RT1 for the department that will make sense to those doing the work. RT1 at the department, segment, or project level is a way to continually discuss achieving and living your organization's core values, without having to say it. Most people don't say they are going to actively talk about the organization's core values and how we meet them—everyone just assumes we do. By continually defining RT1, we bring the core value discussion into an executable format that everyone can support.

Start with the Organization

As Patrick Lencioni (2012) covers expertly in his book *The Advantage,* it is essential to have clarity on the organization and the development of core values and strategic anchors that healthy organizations will work from in order to thrive. What I like most about Lencioni's approach is getting to a limited number of clear and real core values and strategic anchors. More importantly, it is not about creating core values out of thin air but rather going through a process to discover them. Put another way, every organization has inherent values that have developed over time, but often they are not written down or discussed; they are just there. By taking time to discover them and making them real—in words that ring true to those in the organization, where they can say "Yes, this is who we are"—you have the best foundation on which to begin your RT1 discussion.

However, even if you don't have clarity on your core values, which is likely when you have a long list of values (typically more than three to five), the RT1 discussion at the organizational level will help clarify which values are core to your organization and how to consolidate and streamline your communication of them to your organization. To define what RT1 is for your organization, you should accomplish three levels of discussion:

1. Core values
2. RT1 categories
3. RT1 statements

Core Values

For each of your core values, you need to have an RT1 discussion—meaning you must identify how, when, or why the core value is RT1 for the organization. There is no correct answer; it is the discussion and discovery that is important here.

When we look at the core values listed for the *Fortune 500* companies, there is commonality across a large majority of them. The top 10 core values across all these companies are shown in Table 3.1.

I see all these values as very important for a successful organization, but I'm not sure if they are really at the core value level. However, as your organization may have a very similar core value list, this is a great example of how to start your RT1 discussion. Because this should be a discussion, bringing together a diverse group of people from across your organization and stakeholders (internal and external) will provide you with the best insight and value on defining RT1. Obviously, the size and scope of the group will vary based on company size, but it is important that you get sufficient representation from:

- Levels—from the board level to hourly workers, all are important
- Departments—from each key department
- Stakeholders—including key external advisers for their insight and perspective
- Geographic—from different geographic regions, which for a small organization could include several units in the same city

Core value	Percentage of companies
Integrity	52
Customer focus	41
Excellence	32
Teamwork focus	28
Employee focus	22
Innovation	21
Respect	21
Responsibility	17
Safety	17
Diversity	14

Table 3.1 The Top 10 core values across all *Fortune 500* companies.
Source: http://blog.panictank.net/core-values-fortune-500-companies/

The Nominal Group Technique

One of the most effective brainstorming tools that I have used for more than 30 years is the nominal group technique (NGT); it is simple to use, gets maximum participation from everyone in the room, and helps eliminate groupthink or having one person drive the discussion/solutions (Delbecq 1975). The NGT is a four-step process, led by a facilitator, that will allow you to quickly get to a definition of RT1 for each core value. Note that the more core values you have, the longer this can take. You cannot shortcut this step by combining or eliminating core values; you must get through each one. The steps are:

1. Brainstorming
2. Discussing
3. Voting
4. Summarizing

Upon completing the core value NGT workshop, you can then discuss the need to change, consolidate, or eliminate any of your existing core values.

Facilitator

In successful NGT workshops, there are two facilitators who manage the workshop and report the results. Ideally, the facilitators are from the organization but not actively participating in the workshop—they are truly facilitating the discussion.

There is a lead facilitator and a documenting facilitator. The lead facilitator is the one in front of the group, asking the questions, recording short responses, and keeping the discussion moving. This person needs to be good at engaging the group, limiting discussion to one person at a time, and quickly summarizing what he/she hears.

The documenting facilitator is typically in the back of the room on a computer capturing the results of the workshop, including the information presented by the lead facilitator to the group, as well as summarizing the discussion by the group.

Brainstorming

The brainstorming portion of the NGT is quite simple: the lead facilitator asks each of the participants to take five minutes to silently write down their definition of RT1 for one of your core values (see Figure 3.1). There are no good or bad answers—just each person's perspective. As general guidance, I ask the participants to be succinct in their answers and not provide sentence-level detail; answers should be one to three words.

Core value _____

Please describe how you define RT1 for this core value.

_____
_____
_____
_____
_____
_____
_____

Figure 3.1 Core value RT1 brainstorming.

After everyone has had time to put their thoughts on paper, the lead facilitator needs to record the responses for all to see. The only rule during the recording of responses is that there is no discussion. As soon as one participant says, "That was a dumb idea," the other person is less likely to share more. Therefore, the role of the lead facilitator is to go around the room, one by one, and record one item from each participant. I like to use flip charts because they provide an easy way to see and move around as the sessions proceed. Once you get through all participants, start over for their second response, then the third, and so on. For a smooth execution of this phase of the NGT, it is best practice to:

- Use the round-robin approach to collecting responses—only one answer from a person at a time.
- Don't worry about repeats—it is more important to get responses than to consolidate.
- Reply to participants what you are writing—often, you will summarize their response for clarity. It is OK to ask, "When you say X, does this mean Y?"
- If someone does not have anything more to contribute, he/she can say "Pass."
- After a couple of round robins, when you are getting a lot of passes, just open it up for any remaining ideas.
- Number each response to make the discussion and voting go more quickly.

Discussing

Even on a simple topic such as a core value, I find that you can easily have 30 to 50 responses. However, due to the no-talking rule when recording responses, not everyone in the room may clearly understand what each person meant by each response. Therefore, the second step in the NGT is to discuss each response.

The lead facilitator's role in the discussion phase is to play dumb and simply ask the group for each response, "What did you mean by X?" It is not required that the person who provided the item give a response but that the entire group has the same understanding of what X means. The discussion phase is also when you have the opportunity to consolidate items, especially for repeated items. It is important, however, to cross-reference items rather than delete them. For example, if #3 and #15 are determined to be the same, simply put a note saying "See #3" next to #15.

As mentioned previously, during the discussion, the documenting facilitator is in the back of the room recording the items and the discussion. After the brainstorming session, a detailed report of priority items and their meaning can be created easily and quickly.

Voting

Once everyone understands and is aligned on what each response means, the next step in NGT is to have each individual vote for the top five items. To get good results on voting, it is important that the participants use the following process:

1. Items of interest: On a piece of paper, individuals will go through the entire list of items and identify those that interest them (see Figure 3.2). These lists need to have at least five items but can have as many as they feel are relevant. The intent of this step is to reduce the large list to a smaller list they can focus on.

2. Reduce to five: The second step is to take their interest lists and reduce them to five items. Individuals can use whatever process or approach they would like; they just need to reduce to their top five.

3. Rank the top five: The final step is to rank their top five using the following approach:

    a. Choose their first choice.

    b. Choose their fifth choice.

    c. Choose their second choice.

    d. Determine their third and fourth choices in any manner.

1. Please review all responses and list those of interest here (use #s)

_____

_____

_____

2. From the list in question 1, reduce to only five items, no order

_____

3. Rank order your top five items by:
   a. Choose 1st choice
   b. Choose 5th choice
   c. Choose 2nd choice
   d. Determine 3rd and 4th choices

Rank	Item #
1	
2	
3	
4	
5	

Figure 3.2 Core value voting.

This approach to voting has been proven to be much more effective than just rank-ordering five items from the overall list, and it will provide you with sound results.

For determining the overall rank of the entire list, the documenting facilitator tallies up the individual votes, with the first rank item getting five points, the second four points, and so on. Obviously, a spreadsheet can be used to do this quickly.

Summarizing

The final step in the NGT is to summarize the results for the participants by sharing the voting results immediately and, in the longer term, by sharing a report on the workshop.

Beyond the immediate ranking of items, it is important to look at the details of rankings by the various groups that participated. The two that I typically focus on are level in the organization and geography. By taking the rankings and grouping by these two categories, you can quickly identify differences in priorities between the groups.

It is important to specifically look for items where one group ranked the item high and the other group ranked it low or not at all.

So, is one group right and one group wrong? No, as the purpose of the NGT is to get diverse inputs and perspectives. One of the greatest benefits of using the NGT is that you quickly identify these differences between groups, so they can be surfaced, discussed, and understood. Think of it this way: the differences will eventually surface as conflict or animosity in the broader group, as they don't understand why something that is important to them is not important to others.

Through this process of analysis and discovery, you have the opportunity to gain greater alignment and understanding simply by showing the group the results and asking why an item was important or not. I have found it does not change anyone's ranking or the overall ranking, but everyone better understands each other's perspective; in the end, participants become more committed to the overall group's ranking and not an individual's or subgroup's ranking.

Core Value Discussion

The final step in the brainstorming process, after a workshop has been done on each core value, is to take a step back and review all core values again. Often, what you will find is that several of your core values are similar or redundant and can be combined, or you will find that some core values no longer belong or don't truly fit your organization.

To have this discussion, create a simple table that has your core values and the top five RT1 responses (Table 3.2). This provides you with a quick and easy way to see commonality across core values and common definitions. Have the group review and discuss the table and highlight those core values that need changing, consolidating, or eliminating. In addition, you may find that you will need to add a core value that was not identified before.

The goal of this core value discussion with respect to RT1 is that at the end of it, your team has a better understanding and commitment to your core values. In many cases, the words used to verbalize your core values are not changed, but everyone has personalized and internalized them through this process.

RT1 Categories

Once you have alignment and an understanding of how RT1 is defined for your core values, the next step is to identify the key categories of RT1. Simply put, across your core values, you will find commonality with some of the definitions of RT1. They could be the same word or similar in nature.

However, one word of caution: Even if it is the same word, the context among core values may be different—this is where the descriptions of each RT1 come in to provide the proper context. In some cases, you may need to change the RT1 description to better match the context; in others, you may be able to keep the description as-is.

Core value	RT1 1	RT1 2	RT1 3	RT1 4	RT1 5	Change	Combine	Delete

Table 3.2 Your core values and the top five RT1 responses.
Note: The shaded area indicates a likely need to consolidate core values—a maximum of three to five core values is best practice and reasonable.

There are two approaches to identifying/creating your RT1 categories: core value or new category.

- Core value approach: In this approach, after discussion, you find that your core values are very concise, and the RT1 list with each core value flows well and is descriptive of the core value. In this approach, your core value is your "RT1 category," and you are just creating an "RT1 statement" for each core value.

- New category approach: In this approach, after discussion, you find that there are different (often fewer) RT1 categories than there are core values, and it is easier to establish new categories than trying to write a statement for each core value.

If you get stuck on identifying your RT1 categories, a good place to start is to consider the three categories of people, planet, and profit. Though not always a great fit, it does get the discussion going to then identify your final categories.

For each RT1 category, include a list of the RT1s that you identified during the core value exercise, and reorganize if creating new categories. As you look at them, there should be good alignment and a natural understanding of how they go together. In the end, you should have your RT1 categories with a list of RT1s, either directly related to a core value or associated with one or more core values.

For example, Organization A has adopted five core values and has defined its RT1 categories around the people–planet–profit approach, as shown in Figure 3.3.

Core value		RT1 category	RT1 1	RT1 2	RT1 3	RT1 4	RT1 5
Integrity		People	Collaborative	Wellness	Balance	Respected	Aligned
Customer focus		Planet	Once	Regenerate	Circular	Social	Equity
Excellence		Profit	Living	Growth	Value		
Teamwork							
Responsibility							

Figure 3.3 Core values in relation to RT1 categories, for people–planet–profit.

RT1 Statements

Now comes the fun, but often hard, part: actually writing your RT1 statements. The fun part is that this establishes your starting point and anchors for driving Quality First throughout your organization. The hard part is writing concise, actionable, and understandable statements. What I find works best is to have each team member write out a short (one-sentence) RT1 statement for each RT1 category.

These draft statements can be posted on a wall for everyone to look at. It is important to not put names on the drafts, as the point is to have a good group discussion of what they like and don't like about statements, not who created which statement. Typically, at this point, many statements will look and feel similar, and the group is really discussing which statements convey the best message and will resonate with their organization.

One thing to consider when drafting RT1 statements is that the statement can include the list of RT1s or it can be broader in nature. This is more of a preference and style thing for your organization, and both work. For example, using the people RT1 category, you could have any of these RT1 statements:

- RT1 for our people is when we achieve a collaborative environment with alignment on direction and execution of work, where every person is respected for his/her contributions and there is balance and wellness in individuals.
- Our people are our greatest asset and ensuring RT1 with and through them is essential to success.
- RT1 is when every individual is respected; has balance, wellness, and alignment throughout his/her life; and works collaboratively in the organization.

It is important to remember that the RT1 statement is a summary or clarifier of the category and that most of your people will remember the category but not necessarily the statement. Therefore, while the statement is important, it will change and morph over time with continuous improvement, with the categories being fairly fixed over time.

Pilot with Initiatives

Once you have accomplished the work to define RT1 for the organization, it is important to then begin your adoption and use of RT1 by piloting with initiatives, or testing and learning before organization-wide implementation. This needs to be done for two reasons. First, this is new to your organization, and it is important to work through how you will communicate it and use it before expanding across the entire organization. Second, by piloting, you will get the personal testimonials on how it worked to help drive it through others, as well as what may need to be modified for others to understand and do RT1.

The initiative you pick to pilot can be almost anything within the organization. It could be a standard just starting, it could be a pursuit for a new project or work, or it could be an upcoming volunteer opportunity for your team. What is important to accomplish with the initiative is to incorporate the discussion and definition of RT1 for the initiative at its start and the use of RT1 during its implementation. Follow these five steps:

1. Organization RT1 discussion: The first step is to introduce the organization's RT1 statements to the initiative team and have the team discuss them. Ideally, each individual should be able to personalize each RT1 statement to something he/she directly does or is responsible for. This is an important step, in that you can translate RT1 from an organization's perspective to that of the individual (which is important in the next section, where we focus on the individual).

2. Initiative RT1 discussion: Once there is an understanding of the RT1 statements for the organization, the team needs to discuss how those RT1s flow to and through the initiative. This will vary with each initiative, but it is important to have this discussion—primarily because not all RT1s may be applicable or as important in the initiative. By having this discussion, the team can prioritize and understand how to go from the organization level to the initiative level for each RT1 that is relevant.

3. Initiative RT1 statements: The next step is to look at the organization's RT1 statements and determine if new statements need to be made for the initiative. Typically, this is required when the initiative is small in scope and RT1 can be defined as a subset from the organization. In the end, it is

important for the team members to be able to easily say that RT1 for this initiative is X. If they cannot, then they will not be focused on doing RT1.

4. Doing RT1: The best way to do RT1 is for the team to have ongoing discussions; this could be daily, but at least weekly, on what is right and how it is being accomplished. By making this part of and integral to how the team plans and executes the initiative, it becomes a common focus and helps lead implementation.

5. Initiative completion RT1 discussion: The final step is to have the team discuss how defining and doing RT1 affected the initiative. This can be positive, negative, and neutral. For the positive side, it is important to capture the value proposition and personal stories. For the negative, it is important to understand what the impact was and whether it can be eliminated (continuous improvement). The neutral typically happens with very-high-performing teams that already have internalized RT1 without verbalizing it or focusing on it.

My recommendation is to accomplish RT1 pilots with 5 to 10 initiatives and then summarize the findings before proceeding further. By pausing and looking at the results, you will quickly find out if the organization's RT1 statements (and sometimes your core values) resonate across your organization, as well as how focusing on doing RT1 affects your organization. From this pause, you may need to modify the RT1 statements, as well as how you introduce and roll out the statements to your organization.

Focus on the Individual

Once you have finalized the organization's RT1 statements and determined how best to communicate and utilize them, the next step is to focus on the individual. As discussed in the second foundational element, RT1 can only be achieved by the individual. Therefore, before implementing and driving RT1 throughout the organization, just as with the pilot initiatives, it is important to start with the individual.

Because you want RT1 to be personal and visceral with each of your teammates, it is important to approach this carefully and thoughtfully. An email blast from the CEO, though it may be needed, is often impersonal and will not get the desired impact. My best practice approach is to create and implement a "define and do right campaign," which starts with the C-suite and flows out through leadership. This instantly conveys its importance, as it comes from the top down; but by following a simple approach, it also engages the individual in the discussion and personalization, especially the do part:

1. Use pilot initiative leaders. Your strongest and biggest proponents will be your leaders from the pilot initiatives who clearly communicated

value at the end of the initiatives—they have the personal stories of why it is important to define and do RT1. Use these leaders in the C-suite engagement. While this may seem daunting to some, trust their passion and personal stories; it works.

2. Start with the C-suite. Following the pilot initiative, each engagement introduces and discusses the organization's RT1 statements and core values. The primary goal of the C-suite's engagement is to gain an understanding and buy-in to the organization's RT1 statements and the importance of using them in daily planning and implementation. A great approach for this engagement is to have each C-suite member personalize the RT1 statements by identifying how they impact him/her directly in driving his/her areas of responsibility. The personalization of the statements by the CEO, COO, and the like enables them to understand how a statement applies to them, as they apply it to everyone.

3. Engage your level 1 leaders. Following the C-suite engagement, we then move to the level 1 leaders in the company—those individuals who report to the C-suite leaders. For most organizations, by engaging the C-suite and level 1 leaders, you capture the attention of those who affect where the organization is going and how it will get there.

4. Use a one-up approach. As you roll out to all individuals across the company, it is important to use a one-up approach. For example, if there is engagement with a team that reports to a level 1 leader, you would want the level 1 leader plus one level up involved in the engagement, which in this case would be his/her supervisor from the C-suite. This means most leaders would typically be involved in 8 to 12 engagements as they spread through an organization. By using the one-up approach, you get consistency in messaging, but you also get some repetition in the discussion and implementation by leadership, which helps reinforce the definition of RT1 to them as well as helping them continue to use it in their area of responsibility.

5. Incorporate continuous improvement. The final element of engaging individuals is to incorporate continuous improvement in the engagements. Put another way, how the engagement is accomplished the first time with the C-suite should be different, sometimes widely different, from how the engagement is done with a frontline worker. The reason is that at the end of each engagement, we are asking what worked, what did not work, and how we can do better. All of these touchpoints need to be consolidated, and recommendations for improvement must be shared with those doing new engagements.

I am sure you are thinking to yourself, "Wow, this is going to be a lot of work and really hard to do!" I won't sugarcoat this; it is a lot of work. However, think about any large change implemented in your organization and how it was rolled out and how successful it was. I would venture to guess that it took a lot of effort, but it may not have been that successful—likely because the second foundational element of the individual was missed.

By taking your time and putting in the effort on this step of engaging the individual on RT1, you are creating the catalyst and energy to supercharge your entire organization to do RT1. There are no shortcuts to this step, and it must be done, and continually done, as new people join your organization, and it must become part of an integral discussion on everything being accomplished.

Drive Through Departments

The good news is that once the individual engagements have been accomplished, driving RT1 through an organization's departments, segments, and groups becomes relatively simple. In fact, the approach used for the pilot engagements becomes the standard for day-to-day operations, as well as any new initiatives or projects being accomplished:

1. Organization RT1 discussion: It is important to always start with the organization's RT1 statements, as these may change over time.

2. Department/segment/group/project RT1 discussion: In any organization, it is important to tailor RT1 for the department, segment, group, or project so the individuals can directly do RT1. This does not change what RT1 is for an organization but rather puts it into the context of the specific work being accomplished. What is important is that the team is aligned around what RT1 is so they can do RT1.

3. Doing RT1: As with the pilot initiatives, an ongoing discussion of RT1 must be integrated into daily stand-ups, weekly meetings, and the like. By making this integral to how the team plans and executes its work, it becomes a common focus and helps lead implementation.

4. Continuous improvement: Consistently challenging and improving the definition of RT1 will keep the team challenged and improving over time.

Doing RT1

At an organizational level, once you have defined RT1, then doing RT1 requires clear and consistent communication, the use of positive reinforcement, constructive feedback to problems, and integrating doing RT1 into the reward structure of the organization.

Communicating and Positive Reinforcement

Every organization has a style, method, and system for communicating. There are also information communication systems that are often much harder, if not impossible, to influence. One thing you will notice as you advance on your Quality First journey is that how you communicate defining and doing RT1 will change. At the early stages, you will likely be much more direct in your statements and use the terms of defining RT1 and doing RT1.

However, as your Quality First journey matures—and, more importantly, as your definition of RT1 becomes intertwined with your core values—the messaging of RT1 becomes supportive and implied and often less direct. Regardless of where you are on your journey, there must be continuous communication with respect to defining and doing RT1, and it must be included as part of other key communications. For example, when the CEO sends out a business update to the organization, reinforcing how RT1 has affected and improved the company will show clear support and a need to continue its use.

Another key element of doing RT1 and communications is the use of positive reinforcement. Simply put, if someone is doing his/her best to do RT1 and fails, hammering that person with negative comments will not get him/her to do better. Yes, we must learn from failures, but you should not berate people. Instead of asking "How could you screw up so badly?" ask "While we did not achieve RT1 in this case, what can we learn from this and continue to improve?" This is not just semantics, and we are not ignoring the issue or the problem; we are addressing the challenge every organization has—learning from failure and making it a positive, instead of not learning and continuing to repeat mistakes.

The other aspect of positive reinforcement is capturing and communicating successes—both small and large—to the broader organization. Part of doing RT1 is recognizing those who have done RT1. I find that this can be as simple as storytelling, where when you get into groups at company events, each person brings a story of achieving RT1. This takes away the challenge of documenting success by just having people talk about it.

A second way to accomplish more direct positive reinforcement is to publish short articles on success on the organization's intranet. This could be simplified with a short form anyone could fill out highlighting their RT1 and how they achieved it. By collecting these over time, you will have your case studies of the value of RT1 for your organization.

Addressing Problems

As mentioned under positive reinforcement, how you address problems in a Quality First organization sets the tone for how people execute RT1. The goal is full transparency in defining and executing RT1 and that the two match up.

However, organizations and life are complicated, and we do not always achieve RT1. Think of any inventor, like Elon Musk, Nikola Tesla, Leonardo Da Vinci, or Alexander Graham Bell; they all had one thing in common: immense failure. It was in their failures that they found their motivations and solutions. Therefore, a key tenet of Quality First is positively addressing problems. You should:

- Assume the individual last: This is probably the hardest for most managers because they have been conditioned to immediately blame the person when problems occur. However, by assuming anything but the individual caused the problem, you address the problem first and foremost and the individual last.
- Understand: To understand the problem, you must look at the definition of RT1, what knowledge and tools were required to achieve RT1, and what predecessors may have affected not achieving RT1. For large problems, this understanding phase could take days, weeks, or months. But for most, it is a short conversation to identify the glitch in achieving RT1.
- Refine RT1: Many times, once you understand what happened, the definition of RT1 must be changed or refined to align with what is needed and possible. This is part of the continuous improvement foundational element.
- Support doing RT1: Finally, support doing RT1 by recognizing when it is achieved and especially once it is achieved after a failure.

Total Rewards

The last part of doing RT1 is to look at how you recognize and reward people in your organization. If rewards are strictly given based on the money made, then the incentive is to make money, not necessarily to be Quality First. One word of caution: Many organizations base rewards on "performance," which is ill-defined in the organization and often ends up focusing solely on the money.

Therefore, it is important to take a hard look at how you address total rewards within your organization, including how you do performance reviews, how you give bonuses and raises, and how you recognize individuals and teams throughout the year. Using a balanced approach to rewards is required to become a Quality First organization, and it is a key element of incorporating RT1 and continuous improvement into the approach.

RT1 Foundational Element Activity

Now it is your turn to apply what we have covered in this section. Using Figure 3.4, think about your current organization, your department, and your role within the organization, and accomplish the following:

1. Brainstorm how you would define RT1 within the organization (at any level)—there are no wrong answers.
2. Indicate which level within the organization each of your responses applies to, which could be one or more levels.
3. Review the list and identify (or consolidate/create) the top three you believe are essential RT1s that must be focused on within your organization for it to be successful.

❶ Brainstorm definition of RT1	❷ Categorize level RT1 applies to		
	Organization	Department	Individual
1. _____	☐	☐	☐
2. _____	☐	☐	☐
3. _____	☐	☐	☐
4. _____	☐	☐	☐
5. _____	☐	☐	☐
6. _____	☐	☐	☐
7. _____	☐	☐	☐
8. _____	☐	☐	☐
9. _____	☐	☐	☐
10. _____	☐	☐	☐
❸ Personal top three essential RT1s			
1. _____			
2. _____			
3. _____			

Figure 3.4 Summing up the RT1 foundational element.

You have just started your Quality First journey by identifying three focus points of RT1. As you continue through the book, use these three as your focal point in discussions and development. If you are doing this with a group, even better—use the consolidated list to start your journey.

To continue building on your Quality First journey, next, we'll move from defining to doing RT1 with foundational element two—individual/collaboration.

Chapter 4
Foundational Element 2: Individual/Collaboration

The second foundational element—individual/collaboration—is where value is created, as it is only the individual who achieves RT1. Put another way, the individual's work is either correct or it is not, and when it is not, there are issues downstream, often resulting in some form of rework.

For an individual to achieve RT1, that person must understand what RT1 is, have the proper knowledge at the correct time, and know how to verify RT1 when doing his/her work. The constant alignment and knowledge of RT1 is the hardest part, and that's where the collaboration portion of this foundational element comes in. It is only through the collaboration of individuals in planning the work that RT1 is understood and known by all individuals. This includes the upstream handoff from the previous individual and the downstream handoff to the next individual.

It is also through collaboration that the commitment to continuous improvement comes in, as the members of the team hold each other accountable for improving how they execute their work. With continuous improvement, then, comes additional value, as RT1 continually improves along with the execution of work and learning better ways to work.

Individual

Not to discount teamwork and collaboration, which I'll get to in the collaboration section, but I do believe that the hyper-focus on building teams and collaborative workspaces over the past 20 years has taken the focus away from the fact that all work is ultimately accomplished by the individual. But what about a true team activity where you need six people to move an item across the room? OK, let's talk about that. That team has six individuals, and each individual needs to understand what RT1 is for the activity to be successful—which is likely different for each person. In addition to lifting their portion of the load, which could vary from person to person, potential differences in RT1 include:

- The person in front, who can see where to go, will succeed by navigating to the correct spot.
- The person in the middle, who can see a little of where to go, has a blend of following and leading.
- The person in the back, who can only see the box he/she is moving, must explicitly follow the others to move in the right direction.

Now I know, in this simple example, that you probably won't get into this level of detail, but it follows with any activity you do—each individual determines the success of his/her scope of work, regardless of the team arrangement or flow of work. Understanding and truly believing that only the individual can achieve RT1 is sometimes hard for people to understand or get behind—but it is essential that you make this transition, or even a leap of faith if need be, because it affects how you set up and implement Quality First in your organization. When something goes wrong, it is typically because of a series of events that occurred across multiple individuals. This may cause confusion to some because I say RT1 is only achieved by the individual; however, even in complex cases, often a single action sets the issue in motion. Let's look at a few cases, as shown in Figures 4.1 through 4.4.

You walk into a house you want to buy, and in the great room something looks "off"—the main wall seems slanted. Upon further inspection, it is actually leaning 5 degrees, which is not much, but enough.

For this to happen, it is obvious that the person framing the wall had to install it slanted, and that is where the issue started. However, every subsequent person doing work (drywall, painting, light fixtures, etc.) accepted the slant as OK.

We don't know if the initial person was new, maybe the wall was knocked out of alignment later, or maybe the original documents actually had it slanted. Regardless, it was not corrected.

Figure 4.1 The slanted wall.

Unfortunately, in hospitals, there are many issues of giving the wrong medicine to a patient—sometimes the impact can be severe, and hopefully most times it is not.

When looking at the process, there are several individuals involved—the doctor prescribing the medicine, the pharmacist filling the prescription, one or more individuals transporting the prescription, and the nurse administering the medicine.

An issue could be as simple as not being able to read the doctor's handwriting or mislabeling the medicine in the pharmacy. However, with RT1 being no wrong medicines, then each individual along the way must do his/her part to verify that and question when there is a concern.

Figure 4.2 Medicine mishap.

With social media, when something goes wrong, it is immediately caught by someone and rebroadcast around the world. I have seen this especially prevalent with greenwashing, where a company takes credit for being environmentally friendly when it really is not.

If a company defines RT1 as no greenwashing, then each individual in the production and release of publications would have that focus on responsibility to question if he/she is greenwashing or not.

Figure 4.3 Perilous press release.

A very personal risk to many is that of food allergies when eating out. As described in the main text's earlier restaurant and cold food example, there are many individuals in the chain, but the most important is often the server. It is the server who, when taking the order, verifies the allergy and communicates that to the kitchen (and gets confirmation), as well as verifies when the order is picked up that the allergy was addressed, and verbalizes it again when delivering the food to the person at the table.

Figure 4.4 The food allergy.

Empowerment/Expectations

Having the individual as a foundational element brings both empowerment and expectations to each individual throughout the organization. Empowerment is an important concept—it is giving someone the power to do something or take an action under his/her own free will. In very hierarchical organizations, truly empowering employees is a challenge, as leaders and managers do not want to relinquish power—whether perceived or real.

However, the empowerment in Quality First is not a free-for-all; it is maintained around a focus on RT1 and continuous improvement. It is aligning the individual with RT1 and continuous improvement that provides the ability of an organization to first trust and then empower its people, as it enables the individuals to use their skills, creativity, and passion to do their work as best they can and to continually improve how they do it. Put another way, the individual–RT1–continuous improvement triangle provides the context and expectation for how to enable empowerment and make it part of the organization.

It is important to recognize that empowerment will not happen overnight, nor will it sustain itself without reinforcement and focus by leadership. To accomplish this, you can utilize the four foundational elements in a mini-empowerment cycle to jumpstart empowerment and your Quality First journey.

The key elements of the mini-empowerment cycle include:

- Empowerment of RT1: The first step in empowerment is to define what RT1 is relative to empowering your individuals. By going through the 5-W process, you can define what empowerment is and why it is important to your company and develop the empowerment RT1 statement.

- Individual empowerment: Actually implementing empowerment will be organization-specific. In a small organization, it will likely be organic and natural, whereas in a large organization, you will have to work through multiple layers, as well as preconceived and tacit resistance to change. Use empowerment RT1 as the starting point to discuss empowerment with individuals and teams (collaboration), and enable (empower) them to begin the best way for them. The one thing I know is that empowerment cannot be prescriptive; it must become integral to how an organization operates and behaves.
- Verify empowerment: Embedded formal and informal verification of empowerment supports its use and shows value to the broader organization. Storytelling and highlighting successes in casual settings is one of the best ways to verify and drive empowerment. It is important to avoid having verification become a task or a burden, as that actually degrades empowerment in itself.
- Continuously improve empowerment: The last part of the empowerment cycle is to improve it by learning what is working and what is not working, and to address if empowerment RT1 is still valid for the organization.

The cycle time will vary based on the size of the organization and where you are on your Quality First journey; it could range from weeks to months. Leadership must stay focused on driving and improving empowerment throughout the organization.

Knowledge Nexus

One of the biggest challenges, and also the biggest opportunity, in any organization is perfecting the *knowledge nexus*. That means getting the proper information at the correct time to the individual doing the work, so he/she can do it RT1 (Figure 4.5).

Think about the knowledge nexus so you can make your organization a knowledge organization, where the individual naturally uses and creates knowledge as part of his/her work. This is harder than it sounds, as you don't want the focus to be *any* knowledge but rather knowledge that adds to the greater good of the organization—and knowledge is continually improving. The challenge is really that with unlimited information at your fingertips (your phone), there is the misconception that *information* is knowledge.

At the heart of a knowledge organization is a focus on how knowledge is created and used by the organization and by its individuals. To be successful, there must be a clear expectation that as part of your work, you use and create knowledge that is either new or an addition to existing knowledge. In best-in-class knowledge organizations, at the end of each project (or even at the end of each day), staff members meet to formally discuss "knowledge" and their overall contribution to the organization's knowledge.

One important aspect of a knowledge organization is that it is not all about electronic systems and data, as many forms of knowledge flow through the organization. While the electronic systems should help with the capture, organization, and retrieval of knowledge, the organization also needs to tap into the tacit knowledge of its more senior staff members, as well as shared knowledge when working on production lines or similar situations where there is knowledge that flows with the work.

While achieving the knowledge nexus is easier in a knowledge organization, it is essential to focus on the knowledge nexus in any organization to enable the individual to achieve RT1 consistently—the correct information at the correct time to the correct individual.

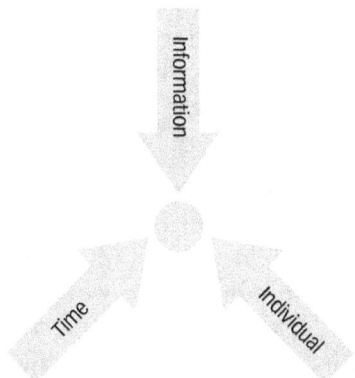

Figure 4.5 The knowledge nexus.

Proper Information

Needing the proper information to do your work right seems obvious; however, let's take a step back and talk more about the process of identifying the proper information in the context of the individual, which will vary based on the task at hand and the person. It is critical to start with the process because an essential element of the proper information is to overcome the innate issue of "experience," where someone has 20 years of experience doing something and that person assumes he/she knows how to do it right. However, we all know that RT1 is continually changing and improving, and if the person is not changing and improving along the way, his/her experience is outdated and will not achieve RT1.

So, let's think about a process to identify the proper information. The process itself has to be flexible enough to extend across industries and complexity of work, as well as be adaptable to changes as work is accomplished. The process includes three primary steps:

1. Define the proper information
2. Package the information
3. Integrate it into the workflow

As there is such variability in work, the first step is to pause and *define the proper information* for the specific scope of work. This is critical, as the proper information can change from individual to individual based solely on their experiences, knowledge, and abilities. Therefore, the key discussion topics on identifying the proper information are:

a. Individual current state: The first discussion point is determining what the individual knows and what his/her capabilities are. If the person has never done the activity before, training and orientation may be required before actually completing the scope of work. This is also where you identify any experience bias, where a person believes he/she knows how to do the work but has never relied on detailed technical documentation before.

b. Information available: There is always more information available than will be consumed by the individual. However, understanding what is available is the first step in identifying what is actually needed. Much of the information flows from the specific work being completed and will vary from a nurse doing a procedure to a construction worker pouring concrete. However, in both cases, there is detailed information available to them.

c. Information needed: To determine the information needed, you must compare the individual's current state with what is available to fill any gaps in knowledge, as well as to align with achieving RT1 for the scope of work. While this is often subjective, the important part is that it is discussed and addressed to narrow the focus on what is needed, not what is available.

d. Noninformation resources: A final discussion point is looking at noninformation resources that may be available, such as mentors, third-party support personnel, and certifications to improve the overall knowledge and performance of the individual.

Once you understand what the proper information is, the next step is to *package the information*. How the information is packaged and provided to the individual is important, because different people absorb information differently. My general guidance on packaging the needed information is to keep it as simple and basic as possible. For example, if pictures can be used instead of text, use pictures. Most people get more out of pictures than text, and a picture gets past potential language barriers. A great approach and a catalyst to Quality First on packaging information is the use of knowledge snippets, which are covered in greater detail in Chapter 10.

Ultimately, information needs to be *integrated into the workflow* itself, where knowledge is created and used as is relevant to the scope of work. This should include the individual in the discussion/creation, but it is just as important to understand the flow of knowledge from person to person when required. A few examples are listed below.

   a. Product approvals: In many instances, a product has to be reviewed and approved by an entity before it is manufactured, installed, or used. In these cases, it is important during the approval process to identify the critical information needed by the individual to properly manufacture/install/use the product. Essentially, how do you pull out the needed information from the entire review package information?

   b. Mock-ups: I'm a big fan of using mock-ups as a way to create the information needed by the individuals doing physical work, primarily because those doing the work create the information they need, in the form they need, as they go through the mock-up process.

   c. Work plans: For scopes of work that are constantly changing, work plans are a great way to determine how the work will be executed through a formal planning process. By integrating the information needed into this planning process, it is not an afterthought but integral to its execution.

One last thought on the proper information: It does not need to be a lot of information, and it could simply be a few lines of text or a picture. During this process, you should continually try to simplify and make the information more digestible by the individual instead of just piling on unnecessary information.

Correct Time

Once you have identified the proper information that is packaged (snippet) and integrated into the workflow, the next challenge is to deliver the information at the correct time. The correct time is at that point in the process where the information will help the individual understand how to accomplish his/her work to achieve RT1 for the scope of work.

In most cases, the correct time will be immediately before the specific task that is being accomplished to get in front of any issues and get the individual to focus on what is important. However, there may be a series of correct times, as there can be several steps in the work process, including setting up to do the work, doing the work, and post-work activities. In other cases, as in production lines, it may just be the one step to continuously focus on.

Similar to the proper information being integrated into workflows, delivering this information at the correct time requires integrating it into your organization's key systems. The best way to think about this is to ask how the information a person needs flows through the existing systems versus having a separate system to manage the key information. Integrating into your existing systems means that the individual only has to go to one place to find the relevant information to do his/her work.

Contrary to this is having your people rely on the internet to find their answers—which, as we know, could be right or wrong. By embedding the knowledge in your systems and processes, the information has been vetted and is relevant to your work, not random or possibly incorrect. However, embedding the knowledge is not easy and must be accomplished strategically and with consistency, which are discussed below.

- Strategic—You cannot get all the knowledge everywhere, as the integration into systems requires time and effort by your organization. To be strategic will require prioritizing what knowledge is the best to integrate and then accomplishing the integration. I recommend reviewing your processes and starting with the one with the highest volume or highest number of touches, which should then have the largest impact on your organization and achieve success throughout your organization.

- Consistent—Driving consistency through your systems with integrated knowledge means maintaining a focus on the knowledge and keeping it up-to-date and improving over time. This is likely the harder part, as you do not want information/knowledge to be stagnant and out of alignment with the current definition of RT1 for the activity. This also helps with continuous improvement and continuing to stretch the knowledge and capabilities of your organization.

Aligning Individual to RT1

As was introduced with RT1, it is essential that the individual is aligned with and understands RT1 for the work he/she is accomplishing. This alignment should be reinforced through the integrated knowledge and use of your systems.

Something to consider from the information and knowledge perspective is to ask yourself if you have the correct person doing the work. While it can be a hard question to answer, having the wrong person will not achieve RT1 in the end. It is not that individuals cannot learn and grow; but at times, they may not have the skills or experience to accomplish the work.

Therefore, as your organization becomes Quality First, you must also address the alignment of RT1 with the individual and the individual with RT1—both must be in alignment over the long term to supercharge your organization.

Collaboration

The second component of this foundational element is that of collaboration and how the individuals work together. While the individual ultimately determines if RT1 is achieved, it is the collaboration between individuals that drives the sequence of work and continuous improvement, and ultimately gets the project completed.

The Concept of Us

As we transition from the individual to the group, I think it is important to briefly discuss the "concept of us," or, more specifically, why is it important to focus on the group and just what is the group. The concept of us, which is different than simply forming a group of people, is where individuals are aligned around the broader RT1 and the process to get there, they are committed to accomplishing the whole, and there is an element of fun in "us."

- Alignment: The focus of RT1 for us is typically one level up from the individual, which could be the project, department, segment, or organization RT1, depending on the context of the work. It is important to gain this alignment, as it provides context to the individual RT1 as well as focus on where he/she needs to go to help achieve it.

- Commitment: When looking at us, the commitment is that each individual will step in to help achieve RT1 for us. For example, if an individual is out sick, other individuals will fill in the gap and keep going. This does not absolve the individual from responsibility but enables fluidity with which the individual will accomplish each task.

- Fun: Call it fun, camaraderie, or something else—within the concept of us, you do not get to us until there is a level of fun beyond the work. While organic in many instances, in others, you may need to create sessions for everyone to get to know each other so they become willing to support each other. Informal and casual interactions are needed to achieve this.

At a minimum, the concept of us needs to achieve alignment and commitment between the individuals and their predecessors and subsequent individuals in their workflow. As people and processes continue to change, so does the need to continually focus on us as well as individuals.

Communicating Through the Chain

One of the biggest challenges with collaboration is how you communicate through the chain of individuals. Ideally, this communication should be intuitive to the group, easy to maintain, and not be hierarchical. In addition, communication should tie directly to the work at hand, as well as be easy to understand vis-à-vis the status of work at any point. This then maintains alignment across the group and also provides a way for all individuals to see how they fit in.

While this sounds too good to be true, there is a simple way to accomplish this—a scrum board. *Scrum*, or "agile," was created through the software industry as a way to plan and execute work in a much more collaborative and ever-changing way than the old linear waterfall scheduling/project management.

In the scrum board concept, an overall project is divided into smaller (typically two weeks) "sprints," where the team focuses on just what is put in the bucket for those two weeks and commits to get the work completed as a team (us). To manage each sprint, a simple graphical board is used to show all the activities to be completed, which are divided into three columns: to do, doing, and done. All activities start in the *to do* column, and an activity is only moved to *doing* when an individual is actually working on the activity. When completed, the activity is moved to the *done* column.

To me, the beauty of the scrum board is that it easily puts into practice both RT1 and the individual, due to the fact that RT1 can be added to the activity card and that only an individual can do each task (or be responsible for completing each task). In addition, by being graphical, anyone can look at the board and immediately see the overall status (how many activities are currently being worked on, how many are done, and how many are still to be completed).

The other value of the scrum board is that as changes occur, they are quickly incorporated into the board, such as new or changed priorities for the team.

Collaboration Tools

Several collaboration tools can be implemented throughout your organization in addition to the scrum board. These include implementing a big room for collaboration, using daily stand-ups for alignment and improvement, and tracking and addressing commitments across the group.

Big Room

The big room is where you bring in different stakeholders on a project and locate them in the same space to work together. This is a very effective way to drive the concept of us, as you can incorporate alignment, commitment, and fun into the layout of the space and how the individuals work together. Some best practice tips on big rooms include:

- Cross-pollination: The individuals from the different stakeholders need to be intermingled with each other. When someone visits the big room, he/she should not be able to know which individual is with which stakeholder—this achieves us.

- Huddle space: Each small group should be arranged in huddle spaces where their workspaces are in close proximity around a central space where they can huddle to discuss work, challenges, and so on. The huddle space is where collaboration occurs.

- Visuals: Throughout the big room, visuals should be prominent to highlight the project, challenges, status, and the like, for the broader team to see what is going on—it could be scrum boards, status graphics, or one-page plans. As you walk around the big room, each huddle space should have something you can look at to see current focus and be able to have ad hoc conversations on what interests you.

- Onboarding: Beyond the physical attributes, the big room needs to have an onboarding process for new individuals entering. This could be simple signage for visitors, or a multiday process for new teammates to acclimatize them to working in a big room.

- Games: To encourage the fun in the concept of us, having game space scattered throughout the big room provides a way for individuals to interact without talking about work—though I find even when playing, work is discussed. If you put in a game, say ping pong tables, encourage use by starting a club or tournament.

Stand-Ups

Whatever the size of your group, one of the best ways to increase collaboration is through the use of daily stand-ups. Within 10 to 15 minutes, all individuals in the group should be able to address:

- What he/she did yesterday and how he/she improved
- What he/she is doing today and how he/she will improve
- What challenges he/she needs help with

It is through these quick hits that you maintain alignment across the group and avoid individuals becoming islands and not getting the help they need. Stand-ups also work extremely well with scrum boards, because individuals can physically move activities to show what they are doing.

Commitments

The last collaboration tool is one of tracking and learning from commitments, which works best when implemented with a scrum board. Essentially, a commitment is me telling the group that I will accomplish something today (an activity). The next morning, if I have completed the activity, I kept my commitment, and if I did not complete it, I did not keep my commitment.

Now, this is very important—if I did not keep my commitment, your reaction must not be negative. You cannot say, "How could you not get that done!?" As soon as you say that, the individual will be less likely to be honest about future commitments. What you do want to do is track the commitments and over time look for trends that need to be discussed. The goal is to learn from not meeting commitments and resolve the real issue, not to beat people up.

Also, a word of caution. Even in the best organizations, keeping commitments typically tops out at about 96 to 98 percent, simply because we work in complex environments and it is hard to be perfect every time. Also, if you are constantly at 100 percent, it is more likely because your staff members are being very conservative in their estimates. This needs to be addressed because it is a sign that you are not continually improving and are not getting the true value from your people and processes.

Individual/Collaboration Foundational Element Activity

It is your turn again to apply what you have learned in this chapter. From the RT1 foundational activity, start by picking one of your top three RT1s and accomplish the following (see Figure 4.6):

1. Document RT1: Write down the RT1 from the previous activity you want to focus on.
2. Identify the primary work accomplished associated with achieving RT1 and provide a brief description of it.
3. Support: Document the key information, workflows, materials, and so on that are needed to do the activity right.
4. Handoff points: Document incoming and outgoing handoff points (other individuals).
5. Collaboration: Identify collaboration elements needed for you to do the activity right.

① List the RT1 priority in Chapter 3 here

_____
_____
_____

② What is the primary activity you accomplished to achieve RT1?

_____
_____
_____
_____

③ Support items

1. _____ 4. _____ 7. _____
2. _____ 5. _____ 8. _____
3. _____ 6. _____ 9. _____

④ Handoff points

In _____
Out _____

⑤ Collaboration points

1. _____
2. _____
3. _____
4. _____

Figure 4.6 Summarizing the individual/collaboration foundational element.

You should now have a good understanding of the individual and collaboratory items that are associated with achieving your priority RT1 identified in the previous chapter's activity. Take a little time to review this, and even better, discuss it with your team on their understanding of the points of connection and their roles in achieving RT1 relative to the activity.

To continue building upon the foundational elements, next we'll examine foundational element three—embedded verification.

Chapter 5
Foundational Element 3: Embedded Verification

Embedded verification is establishing a process where the individual doing the work (foundational element 2) can understand and verify that he/she did the work right (foundational element 1). The focus is on understanding and doing RT1, and then documenting that it was done RT1—which is the verification step. The whole goal of embedded verification is to eliminate all rework—to be RT1 every time.

The biggest challenge with embedded verification is truly making it embedded in your organization, whereby the act of verification is accomplished as part of the workflow and not a separate activity. The general guidance on how to embed verification is to make it as simple as possible—do it with a simple question or a verification checkmark. The more complex you make it, the less likely it will be accomplished effectively.

Importance to Quality First

Embedded verification, when done properly, is how you get the individual to understand and do RT1 before he/she does the work, as well as identify how to continuously improve what the individual is doing. Therefore, relative to Quality First, embedded verification is where the value is created because it is at the point of executing the work.

Integrating the verification function throughout the organization brings several value propositions through Quality First:

- Focus RT1 at the point of need: When properly accomplished, embedded verification reinforces RT1 just before the work is accomplished, enabling the individual to remember what RT1 is and how to accomplish it.

- Data to improve: When verification is integrated into and through your systems, data are available that clearly show performance, both at a point in time and over time. This enables focusing on improvement and also visually showing improvement throughout the organization.
- Individual ownership: By having the individuals own the verification of their work, they take accountability for doing it right. By having this come from Quality First, it becomes the cultural norm within the organization and the expectation of behavior.

Verification versus Inspection

Before we get into the details of embedded verification, it is important to briefly discuss the difference between verification and inspection. To me, verification is the act of verifying that the work is completed properly while the work is being accomplished, by those completing the work.

Conversely, inspection is typically done by a person who did not do the work as a validation step to ensure that all the work is correct. While inspection is and can be of value and is often required as a back-check, it is important to recognize that at the point of inspection, the work has already been accomplished by an individual and is either right or wrong. When it is wrong, inspection just results in rework, whereas the focus of verification is to avoid rework.

Is It Just Semantics?

Whenever I bring up the topic of verification versus inspection, a surprisingly large number of people say it is just semantics—that they are both the same thing. I disagree. The distinct difference to me is that there is the *process of verification* and there is the *act of inspection*:

- Process of verification: When you look at embedded verification, it is not a single activity but a process of how to integrate verification into the workflow and the activities being accomplished to enable the individuals to do their work RT1. This means that the process starts well before the work being accomplished and goes beyond the work being done. The use of a checklist, or the individuals documenting that they did it right, is just one activity in the process.
- Act of inspection: While you can argue that inspection is a process, when you look at its core implementation, it is accomplished at a fixed point in time *after the work is completed*. It is an act accomplished by an individual not involved in actually doing the work, with the intent to validate that the work was done correctly. However, in many cases, the perspective

and personal focus of the inspector are to find problems, which is not necessary if it was RT1—or they have their own definition of RT1.

So, in embedded verification, verification is the process of supporting the individual in doing the work RT1 the first time, and inspection is a post-work activity to identify if RT1 was not achieved. Again, there is value and a need for inspection—I do not discount its value, but for the purposes of Quality First, the primary focus is on embedded verification, as it is the primary value proposition throughout the organization.

Inspection's Degradation of Quality

One of the risks I have seen in instances that have heavy levels of inspection is that quality actually degrades over time, compared with similar projects with lower levels of inspection. When analyzing projects, individuals, and their behavior, it really comes down to where inspectors are focusing their efforts.

I have found that at a higher level of inspection, there is a shift in focus by the team from doing the work to managing the inspections. Put another way, they are more worried about the inspection itself and fixing the issues raised from the inspection than doing the work properly the first time. When you think about it, this is basic human nature. When someone else is responsible for checking the work and finding issues, you focus on what you can do but not necessarily on doing it RT1. In addition, given that you now must do the rework, you have less time to focus on your current scope of work.

To avoid degradation and relinquishing of quality to the inspection side, it simply requires a verification-first approach—you hold the individual responsible for verifying that his/her work is RT1. Then the inspection is a formality of checking your process, not just finding issues in the work. When issues are identified, the focus goes back to the process of verification to improve it to avoid future issues.

Embedding Verification

There are multiple ways to embed verification in your organization, and it will vary depending on the scope of work being accomplished and the department you are discussing. For example, how you embed verification in a production line is different than when looking at human resources talent acquisition elements. At a high level, there are three interface points where verification can be injected into your current processes:

- Planning—As part of your planning process for a specific scope of work, it is essential to get alignment on what RT1 is and to document both how RT1 will be communicated to the individual doing the work and how the individual will document he/she did the work RT1 (the verification part).

- Orientation—While the level and type of orientation will change, the important thing is that some level of orientation is accomplished to ensure that the individual doing the work understands and acknowledges what RT1 is and how he/she will document it. This should be done at a minimum on a daily basis, to address any change in the definition of RT1, and after an issue is identified and resolved.
- Verification—Using the developed approach, the individual then documents that he/she accomplished the work RT1.

As with RT1 before, a critical part of embedding verification is to integrate the act of verification into your organization's systems for accomplishing and tracking your work. For example, if we look at RT1 being the well-being of your employees, embedding verification in your systems could include:

- Recruiting: In the recruiting process, include questions concerning wellness, how they could contribute to it, and how the organization drives it.
- Expense reports: In the system used for processing expense reports, highlight the need for quick payment (wellness) whenever questions arise about not paying an expense. Can they call to resolve a problem more quickly instead of using the system or emails?
- Timesheets: Have the system remind individuals with high paid-time-off balances that they need to maintain wellness.
- Performance planning: Have a mandatory discussion on wellness during any performance planning interaction.

By integrating the item (action) that is needed to achieve RT1 into your systems, as the individual completes the activity (system), the verification is captured by them doing standard work.

Using a Quiz Approach

As you start implementing an embedded verification approach into your organization, you can apply a great concept from your school days: a quiz approach. When you were a kid, teachers often gave a series of homework assignments and quizzes before the big test or the end-of-semester final exam. Why did they do that? The homework and quizzes were to make sure you knew the material along the way and helped prepare you for the final test.

So, when we apply the quiz approach to something complicated, the final test would be performance testing a system, a contract-required test, or a test completed by a third party—essentially an inspection activity. For each of these final tests, just as in school, the intent is to score 100 percent the first time. However, without a quiz approach, this does not happen.

Acing Every Test

Similar to being in school, preparation enables you to pass every test you take. A key to this is setting up quizzes to do along the way before the final test. How this is accomplished will vary based on the item you are focused on and the complexity of the final test. For simple items or processes, there may only be one or two quizzes along the way; for complicated items and processes, there could easily be a dozen or more quizzes (see Figure 5.1). In developing the quizzes, consider:

- Supportive: Does the quiz support the final test in ensuring that by passing the quiz, you are likely to pass the test? For example, when you have an item with layers—where after the next layer is added, you cannot go back and access the previous layer—can you test the integrity of the first layer before adding subsequent layers?
- Simple: As the final tests are often more complicated and sometimes require specialized equipment and certifications to pass, you should create the quizzes to be simple yet effective.
- Early: Try to identify quizzes that can be accomplished as early as possible in the process. Often, the focus is on later tests that could be done earlier.

Building enclosures are complex assemblies with sheeting, air/water barriers, openings (windows and doors), flashing, sill pans, sealants, and a bunch of other items. The final test is a negative air water test to challenge the assembly to real world wind and rain conditions. A successful quiz approach has been:

- Water Barrier Quiz—use water spray rack on water barrier looking for leaks.
- Bubble Gun Quiz—use an air bubble gun to verify air barrier, especially at seams.
- Sill Pan Quiz—fill the sill pan with water to identify water leakage.
- Interstitial Space Quiz—pour water into interstitial space (e.g., between brick and sheathing) to verify water drains and does not enter building.
- Sealant Pull Quiz—verify the adherence of the installed sealants.

By taking these quizzes on an ongoing basis, we have consistently passed the final tests, which was very rare in the industry prior to the quiz approach.

Figure 5.1 Building enclosure sample quizzes.

- Creative: You need to be creative, as the quizzes will not be industry-standard tests—those are the final tests. However, by looking at the final test, you can identify key elements that can be broken out into the quizzes.

Approach to Quiz Development

The most effective way to develop quizzes is to tap into the knowledge and creativity of your teams. Fundamentally, individuals do not want to fail and hate to do rework, but they often are not engaged enough in the process to enable them to succeed. This is where quizzes can come in to help them through their work and ensure it is RT1. A simple five-step process can be followed to develop the required quizzes:

1. Start with the final test—I am always surprised how many people working on an item have never been exposed to how the item will be evaluated at completion. Therefore, it is essential to bring together all individuals working on the item and review what the final test is and how it is accomplished. This provides the foundation to understand when and how to do the quizzes.

2. Brainstorm quizzes—Once the final test is understood, have the individuals brainstorm potential quizzes. My recommendation is to use the nominal group technique (NGT) introduced during the RT1 foundational element, Chapter 3, as it enables gaining input from all involved. This provides the most relevant and creative solutions.

3. Try the quizzes—As in most cases, the quiz approach, as well as the quizzes themselves, are new, so it is essential that you try them in practice before rolling them out on a broad scale. Trying the quizzes is intended to determine what works and what does not work and to finalize the approach and guidance for each quiz, including how often it should be taken.

4. Evaluate results—After all quizzes have been refined, it is important to pull together the original group to review the results and finalize which quizzes will be implemented, because not all of them may provide value for the effort required.

5. Implement quiz/test loop—The last step is ongoing. After each final test, the quiz and final test results are reviewed to identify opportunities for improvement, as well as the need to modify a quiz to identify and eliminate any earlier final test failures.

Challenge of Checklists

I'll be honest with you—I have a love/hate relationship with checklists. On one hand, I understand and love the formality of a checklist in standardizing what is looked at and documented at each step in a process. On the other hand, I've had too many experiences where thousands of checklists were created and completed, yet the quality was not improved. Thus, the challenge of checklists.

Therefore, I have constantly battled about the need for and use of checklists in a process. When the process is highly structured and repeatable, they can work wonderfully, but not always. In focusing on why checklists fail, even in the best organizations, I have identified three root causes:

- Relevance: Probably the biggest failure of checklists is their relevance to the scope of work they are supposed to be supporting. When "checklists are required," the default is often, "give me a checklist to fill out." However, giving someone a generic checklist is of little value if it does not align with or help the individual understand and do RT1.

- Engagement: The second-biggest failure is the engagement of the individual filling out the checklist. In many instances, the checklist is not completed by the person doing the work but by a supervisor or inspector; as we learned earlier in the book, this is not embedded verification, and it results in identifying rework, not avoiding it.

- Format: The final failure is that checklists are often in a traditional table format with a bunch of yes/no questions or values to fill out. However, when you think about our multicultural and multilingual workforce, having a single-, or even dual-, language checklist is not effective.

Therefore, most checklists do not cover the first foundational element of RT1 or the second of the individual—they are created out of context and without regard for how the individual will use them.

By taking an embedded verification approach, a checklist is only good if it has been created by and for the individual who needs to use it and focuses on him/her understanding and doing RT1 for the scope of work. Therefore, checklist creation and use need to be embedded in the planning and execution process for them to work—you need to work with the individuals doing the work to create the checklist they will use. Consider the following:

- Verifies RT1—Once the individual understands RT1, he/she needs to translate that understanding to his/her workplan and how he/she will verify right throughout the scope of work. This typically ranges from one to five items to verify—once you go beyond five, you get into minutiae and not RT1.

- Tiered structure—By collaboratively creating the checklist across multiple individuals, the verification of RT1 flows through the work from individual to individual. For complex systems and assemblies, this enables the checklist to be tiered to follow the workflow—from simple to complex. This also enables integration of the quizzes into the process as part of the work, not a separate activity.
- Visual format—A picture is worth a thousand words. By showing a series of pictures of RT1, you eliminate language barriers, and anyone can verify RT1 as they do their work.
- Displays results—The final element is how you show the results of completing the various checklists. Instead of just tables of numbers, be creative and include visual elements, such as icons, colors, or other items that convey success, roadblocks, or focus areas. You want action taken, not just someone reading a bunch of numbers.

Addressing Issues

The reality, for any process and individual, is that perfection is impossible to achieve all the time. Therefore, it is essential, as part of Quality First, for you to establish how your organization positively responds to and learns from failures.

When a Failure Is Found

When a failure occurs, whether small or large, the first thing to do is to stop, take a deep breath, and start with a simple thought—somewhere along the way, the process failed, not the individuals. I know that for many managers and leaders this is probably the hardest first step to take; however, it is the most important. As discussed previously, it has been rare in my experience that an individual deliberately intended to fail. In the vast majority of instances, that individual did not understand RT1, did not have the information or knowledge to achieve RT1, or did not know how to verify that he/she achieved RT1.

When a failure is found, your primary goal is to learn from the failure and implement improvements in the process to keep it from happening again. For small failures, this could simply be a tweak to embedded verification, training, or orientation of individuals on technical details, or the use of a new tool or technique. However, for major/systemic failures, you will likely need to dive deeper into what is really happening to fundamentally change your overall process.

The Root Cause

Quality and management programs have many tools and techniques for understanding failures and how to avoid them. In general, they all are intended to identify the root cause of the failure—the thing that ultimately cascaded to become the failure. There are multiple books written and courses you can take on root cause analysis; a few of the key tools you can implement include:

- Fishbone diagram: The fishbone diagram dissects the failure into different categories to identify potential items that exacerbated the failure, as there is often more than one. Typical categories you can start with include people, machines, methods, materials, the environment, and measurement. For each category, you usually identify three or more exacerbating factors to the failure. I like using the fishbone diagram to flesh out the various factors, which can then be applied to a 5 Whys discussion.

- The 5 Whys: The purpose of the 5 Whys is to "peel back the onion" on the failure. You ask why the failure occurred, then ask why it happened, and so forth. By asking five (or more) "Whys," you will get to the true root cause. I personally like the 5 Whys approach, as it quickly gets past "the individual messed up" to understanding the underlying cause to the failure.

- Pareto chart: When you have data across multiple failures, you can use a Pareto chart to graphically show the severity of failures by different categories (root causes). The Pareto chart helps you focus your broader intervention initiatives to address systemic issues across failures.

- Scatter diagram: When you begin to have large data sets on failures—for example, the number and cost of rework by a work category—you can use a scatter diagram to visualize your high-risk items (high number and high dollar). To truly identify the systemic risk to your organization, split your scatter diagram into four quadrants—the 50th percentile by number and cost, with the upper-right quadrant being those items you need to focus on.

- Failure mode and effects analysis (FMEA): This type of analysis is more of a failure prevention tool used during design to address potential failures or defects that could occur throughout your process and then design these failures out of the product or process.

Regardless of the tools used, the key is that you learn from the failure and make a change.

Continually Refining RT1

The last item to consider when addressing failures is to understand that your definition or understanding of RT1 may need to be improved. Actually, even without failures, you should be constantly refining and improving your definition of RT1, as you can always improve.

Your primary tool for this will likely be the daily stand-up, as it is the point of accomplishing continuous improvement. We'll cover more details on this when discussing the next foundational element.

Embedded Verification Foundational Element Activity

Using the information from the individual/collaboration foundational element activity in the previous chapter, complete the following around embedded verification (see Figure 5.2):

1. Restate the activity—list the activities you hope to accomplish.

2. Review RT1—review your definition of RT1 and identify the high-level approach to documenting that you did it right (for example, checklist, system, or measurement).

3. Identify embedded verification items—document the actual items you will verify.

From this activity, you should have a discussion with your team on how best to implement embedded verification as an individual and collaboratively across the team so that RT1 is achieved and documented appropriately. (Remember to keep it as simple as possible.)

❶ Restate the activity you are accomplishing
_____   _____   _____
❷ Detail your high-level approach to documenting RT1
_____   _____
❸ Identify your embedded verification items
1. _____   2. _____   3. _____   4. _____   5. _____

Figure 5.2 Summary of embedded verification.

Chapter 6
Foundational Element 4: Continuous Improvement

Throughout this book's discussions of the first three foundational elements, continuous improvement has been mentioned relative to its importance in achieving not just each foundational element but also Quality First. The challenge most organizations face is that they want to try to continually improve, but there is often a strong built-in resistance to change by individuals and the organization. Therefore, the key is to focus first on improvement and then on how to make it continuous.

Importance to Quality First

When you think about Quality First coming from your core and emanating throughout your organization, a key way to drive and increase value is with continuous improvement. I liken continuous improvement to individual investing, where you save a small amount of money each week, and over your lifetime, the value increases upon itself through reinvestment and overall growth, to the point where you have more than enough money to enjoy retirement.

With continuous improvement, you make small changes each week so things are easier to do, which creates more value in workflows, and ultimately culminates in significant gains, not just for the organization but also for every individual.

Layers of Continuous Improvement

A common misconception about continuous improvement is that it is the responsibility of others who look at and improve overall organizational processes. But in reality, there are many different layers/levels of continuous improvement that can be focused on and implemented within an organization.

The Layer Cake of Improvement

I like to think of improvement as a layer cake (Figure 6.1), where each layer is a way to improve within the organization, and the layers build upon and are complementary to each other. The four layers are process, equipment, materials, and individuals.

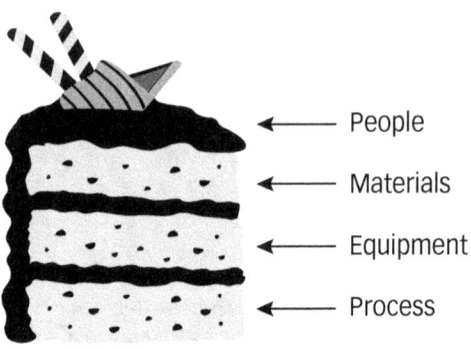

Figure 6.1 The layer cake of continuous improvements.

1. The first layer is *process*. At the base of improvement is a focus on process, where efforts are made to look at a process and streamline it to reduce waste or inefficiencies. This is probably the most classic focus for improvement, and a whole industry has been created to apply workshops, tools, and support for process improvement.

2. The second layer of the cake focuses on improving *equipment*, or the tools used to accomplish the work. This can be physical equipment or electronic tools, such as software or measurements. While this is often part of process improvement, it is important to look at it separately for improvement, given that it provides an understanding of how equipment and tools directly affect—positively or negatively—how your process is working, as well as the individuals in the process.

3. The third layer is the *materials* used within the process. While most think of manufacturing processes, materials are actually much broader, as they are the energy and water used, the facilities you are in, as well as the paper printed by your human resources specialist. This is an important layer due to its breadth, with often unforeseen effects and opportunities on your processes and people.

4. The fourth and final layer is the frosting; it's the *individuals* who hold everything together and are affected by the other three layers. As discussed earlier in the book, individuals improve or will improve the other layers and are essential for achieving continuous improvement in your organization.

Aligning the Different Layers

There are unlimited ways to align and focus on the different layers of continuous improvement, and none are right or wrong, as you can improve in many ways. However, to get the most impact from your investment of time and resources, it is important that you understand and focus on your alignment and make conscious decisions on how to maintain it.

Ultimately, the goal is to find balance in your efforts to improve so they are not skewed to just one area, such as solely looking for process improvements. One way to achieve balance is to look at your improvement initiatives and evaluate how much time or how many initiatives are focused on each of the layers, with each initiative put into just one primary category. This provides an understanding of where your energy is going at a high level and if you have parity across your improvement efforts.

This is not to say that your material improvement initiatives do not involve processes or individuals, but that there are initiatives that are primarily focused on materials so they are not forgotten. As there is crossover between layers, this categorization is very subjective. However, it is important to accomplish, because it provides a good indicator of the health of the overall improvement programs within your organization. I have found that when improvement programs skew to one layer, such as 90 percent focused on process, the program becomes stale and often fails, as the other areas become less engaged.

(Small) Failure Is Good

An essential element of improvement is that not all improvements work or are good for the organization. Some of the biggest process improvements related to changing software have resulted in making things worse for individuals and the organization. Obviously, we want to avoid large failures, but it is important to understand that failure is good—on a small scale, that is.

However, I differentiate between the act of failure (which is bad) and learning from and improving from the failure (which is good). Learning from failure is not a new concept, and it is called different things by different industries. My favorite concept is one of "fail forward"—when you fail, you learn from it and advance from the change and do not move back. Put another way, this is not a two-steps-forward, one-step-back mentality—it is a one-step-forward and then three-steps-forward one, as the failure did not derail us, and we learned from it and then moved even further forward (Maxwell 2000).

For small failures to be good and healthy for your organization, it is important to continue the concept of positive feedback and reinforcement discussed previously when addressing failures in achieving RT1. This includes a learn–implement–check approach to small failures:

- Learn from the failure. The first step is to learn from the small failure to understand what happened or what did not happen. Using root cause analysis tools or simply discussing the failure is important to understand what happened.
- Implement the new change. Failure is good when you learn from it and make a change to keep it from happening again. Therefore, it is important to implement a change to make the failure successful.
- Check for success. The final step is to check for success—that the change you implemented actually worked. If it failed, you need to learn from the new failure and start the cycle again.

It is hard to internalize this and make it safe for everyone to be comfortable, willing, and able to highlight their failures and implement change. I find that the smaller increments in change you make, the smaller failure focus you have, and the easier it is, primarily because people don't get stressed over small failures or talking about them, whereas they have a hard time discussing large failures.

One last thing on failures—you may want to reframe what you consider to be failures to make the adoption of continuous improvement easier. Often, management sees an improvement as a failure if it did not meet the expected improvement, even though improvement did occur. It is important to learn from what did not work, but it's just as important to focus on what *did* improve, so the effort and trajectory are improving. For this to be successful, it is more about how leadership behaves and reacts to the problems (it needs to be positive), as well as how they recognize the improvement made (it also needs to be positive and vocal).

Individual Constant Improvement

As improvement occurs at the individual level, it is important to create a program and focus area within the organization that encourages and enables individual constant improvement. This is a challenge, for it needs to be daily, constant, and ongoing over a long period. To accomplish this and maintain constant improvement requires a blend of the individual and the group—to make it personal and to enable accountability.

Daily Introspection—Individual

To make improvement constant, the individual must take ownership of his/her own improvement, and it needs to become behavioral so it's part of the individual's everyday routine. How this is accomplished will vary by organization and industry, as it can range from direct live visual feedback on a manufacturing line to more ad hoc use of mock-ups in construction. One tool I have found that works well for many is that of a simple daily introspection by the individual.

This daily introspection is not complicated or even deep, as it is intended to address the small improvements and challenges individuals encounter each day. By focusing on small, and not wholesale, change, this introspection can be quickly answered with a few questions:

- *Did the improvement I tried yesterday work? Why or why not?* Since this is a daily activity, starting with the focus from yesterday and why it worked or did not work informs the individual of what he/she should try to improve today. Also, when it did not work, learning from it (small failures) is important.
- *What bugs me?* This introspection is intended to identify where to focus the small improvements. Maybe you're bothered by how hard it is to communicate with someone or that you get too much junk mail. These are not large problems; they are indicators of opportunities to improve.
- *What will I improve today?* With the first two questions answered, the individual can identify what to improve today, which could be a continuation of the previous day(s) or something new. For instance, it could be requesting new signage or it could be setting up rules to automatically delete unwanted mail.

This introspection should not take more than a couple of minutes. You are not trying to solve the world's problems; you just want to make incremental improvements each day. To make this an individual focus and to understand how small changes lead to big improvements over a longer time period, you can integrate the daily introspection into your systems (daily tasks) or use a simple spreadsheet. The goal is to make it easy to accomplish while capturing the individual's improvement journey, which can then be discussed during any performance reviews (see Figure 6.2).

Even if not captured formally, the expectation and act of daily introspection provides value to the individual, as it empowers him/her to improve, and to the organization, as small improvements across all employees result in huge savings.

Documenting an individual's daily introspection and results provides the individual with a great history of improvement and change over time. This is important, as the impact of small daily changes is often hard to see; thus, there is a need to look across a longer period.

Date	Yesterday's improvement		What is today's improvement?	Notes
	Did it work?	Lessons learned		

Figure 6.2 Individual introspection worksheet.

Daily Stand-Ups—Individual/Group

Critical to improving the individual daily is having accountability for improvements. However, this is not top-down accountability from a supervisor, saying improve or else. This is establishing and nurturing collaborative accountability, where the other individuals around him/her are there to offer help and support, as well as to provide critical feedback and guidance to continually improve.

As discussed previously, the best way I have found to create this collaborative accountability is through the use of daily stand-ups, where the small group gathers at the start of the workday and has an informal discussion of the day. This can be an in-person stand-up, with coffee in hand, or it can be virtual—the keys being that they are accomplished daily and that everyone participates. Best practices for daily stand-ups include:

- Have a five-minute casual conversation before the starting time, for those who want to join.
- Give each person 30 seconds to address his/her improvement (from introspection)
  - What I did yesterday—did it work—what did I learn
  - What I am doing today—work focus
  - What I am improving today
- Minimize clarifications or discussion—get through everyone with little, or no, discussion so everyone hears from everyone. This keeps the conversation moving and avoids tangents in discussions.

- Offer closing thoughts. Pick a different person each day to give a closing thought—something he/she heard and thought was interesting or that he/she will follow up on.

To keep the daily stand-ups fresh, it is good to have different people lead them each day/week and to occasionally insert additional questions, such as "What is your favorite local restaurant?" and "What are you doing this weekend?" This helps build team cohesion and camaraderie as you do the daily stand-ups. This is also a great opportunity to give department or company updates.

The biggest value, beyond the daily improvements, that these stand-ups provide is a better, and quicker, understanding of what team members are working on and how they can help each other. I have found that daily stand-ups eliminate the need for many other meetings because individuals connect after the stand-up, and the team overall has a better pulse on the work being accomplished and thus does not need additional meetings or endless emails.

Periodic Pauses—Group

The final piece of individual improvement is to accomplish periodic pauses—take time to discuss and understand what has happened over a period of time. While most organizations accomplish individual-level performance planning and reviews, I find that these don't adequately address continuous improvement and that they are very subjective in quality based on the leader.

Therefore, you should tie the periodic pauses to daily stand-ups so the pause becomes a group activity that spans the individuals and the group they are in. The ideal time span between each pause varies based on the size of the group, the consistency of the daily stand-ups, and group preference. My recommendation is that you start them on a quarterly basis and modify them based on group feedback.

To make the most out of your periodic pauses, the same group that meets for the daily stand-ups should meet for the pause. Here is general guidance on the pause (see Figure 6.3):

- It should be casual/offsite. Given that a primary intent of the pause is to get your team to open up and have a good conversation, it is important to keep the tone of the pause casual, and I would even recommend you do it offsite or at least in a comfortable part of the workplace. As most stand-up teams are 15 or fewer people, a lunch pause can be very effective.
- Start with the stand-up. For the stand-up to be successful over a long period, the way it is accomplished needs to be addressed and can always be improved. Therefore, start your pause by discussing what is working and not working with the stand-up sessions themselves.

This worksheet is to be used by individuals to prepare for a periodic pause as well as to document key items discussed during the pause.

Pre-pause—individual reflection: Please identify the top five items you see as your biggest accomplishments as an individual since the last pause. Then identify the top five opportunities you have to improve in the coming months.

Item	Biggest accomplishment	Improvement opportunity
1		
2		
3		
4		
5		

Pre-pause—stand-up reflection: Please identify what you like about the daily stand-ups and what you think could be improved about them.

Item	Daily stand-up plusses	Daily stand-up improvements
1		
2		
3		
4		
5		

Pause workshop—Right discussion: Use this space to write notes on your team's right discussion.

_____
_____
_____
_____
_____

Pause workshop—actions: Use this space to write notes on any actions to be taken.

_____
_____
_____
_____
_____

Figure 6.3 Periodic pause worksheet.

- Include individual reflection. Take time (this can be before the pause) to allow the individuals to identify what they see as their biggest improvements since the last pause, as well as their biggest opportunities for the future. Each individual should present to the team his/her improvement, with one person providing an overall summary of group improvement.

- Update RT1. With the entire team there, this is an ideal time to review your definition of RT1 for the team, as well as any changes to make to it as the team's members have improved their performance since the last pause. The pause also gives you the opportunity to recommit to RT1 and to continue to stretch what RT1 means over time.

- Discuss benefits. Talk about how the team has benefited from the stand-ups and focus on the foundational elements—RT1, individual/collaboration, embedded verification, and continuous improvement. This discussion is intended to keep the "everyone benefits" discussion going from the individual to the group to the organizational level.

During the pause, it is important to have someone other than the facilitator take notes on key items raised and any decisions made on changing/improving things. The primary reason for this is to ensure the facilitator focuses on having every individual contribute and to actually have discussions—being distracted by taking notes will slow the process and keep you from achieving what you want from the pause.

Process Improvement

Given that there are thousands of books out there on how to achieve process improvement, my purpose here is not to repeat them but to put process improvement into the context of Quality First and how to accomplish it effectively throughout your organization. The three relevant items are where to focus the process improvement, how to organize your process improvement teams, and how focusing on process improvement can supercharge your transition to Quality First.

Team, Department, and Organization

While it is possible to apply process improvement to almost anything within your organization, when process improvement is viewed in concert with the constant individual/group improvement using daily stand-ups and periodic pauses, there are really only three levels on which you should be concentrating your process improvement efforts: team, department, and organization.

Team

From a process improvement perspective, starting at the team level is best, as it is small enough to understand and get your hands around, yet it can make a significant impact on improving how the organization operates. I am using the term "team" loosely, as each organization has its own term for how it executes its primary work/deliverable/project. A few of the key tools to use in understanding and improving your team include:

- A process map: Clearly understanding how work is accomplished and by whom is an essential first step, which can be accomplished by mapping out your processes for the team. These processes typically include both physical and virtual activities (sending information). Depending on the complexity of the processes you are trying to improve, I generally start with a fairly high-level map and then add details as the improvement initiative progresses. This typically saves time in the process as you focus on understanding and not on excruciating details.

- Key performance indicators (KPIs): In addition to the process map, it is important to understand how the team is evaluated currently using KPIs. Unfortunately, for many teams and their processes, they are not evaluated or tracked, and there may be no formal indicators. In these cases, look for informal indicators in the form of spreadsheets or other tracking tools outside normal organizational systems.

- Flow: There are many terms for best-in-class processes. The term I like is "flow," which reminds me of a river flowing past you—it is easy to see and relatively calm to watch. For your processes, you need to discuss how well it is flowing and how can we improve its flow. Looking at wasted time/materials and energy, time of each step, and other traditional process improvement items are possible ways to improve flow.

- One-page action plans: To avoid forming a team to improve your process that never actually gets to improving the process, I recommend that you create a simple one-page action plan for improvement. By keeping it to one page, you must be concise and clear about what needs to be improved and how to do it. In addition, the one-page format simplifies communications to leaders who may have to approve of the changes and avoids long documents that do not get implemented.

Department

The next level of process improvement focuses on the key departments (segments) within the organization and how the departments themselves operate. While there is often overlap between teams and department improvement, departments make an impact on the entire organization and thus can have a larger effect on how things get done.

While the same tools should be applied at the department level that were applied for teams, there is a nuance on what to focus on at a department level. Much of the work within a department is often done by procedure and compartmentalized, meaning that there is some level of repetition. With repetition comes complacency and the notion that this is "just the way we do it."

Put another way, individuals within the department tend to focus on getting the work done and don't see the need or opportunity for improvement. To overcome this, I recommend that departments track the work they do in two separate buckets:

- Planned work—work that has been scheduled, is known, and is accomplished through standard processes or approaches.
- Unplanned work—work that comes up and is not known before and has to be done in addition to the planned work. Unplanned work often has short deadlines and sometimes vague requirements.

While this may seem pretty easy to do and maybe not of value to some, what you will find in reality are interesting results. Primarily, you will likely be surprised by how much time and energy is spent on unplanned work, and that most of your issues and problems are due to unplanned work (see Figure 6.4)

This is not to say that you simply don't do unplanned work—but you must address your (and others') processes to move the unplanned work to become planned work. A simple yet effective way that unplanned work has been addressed in information technology (IT) departments is by the establishment of service-level agreements, where typical requests have typical response times associated with them.

Department: Marketing

Unplanned work: Requests by managers for initial materials to respond to customer opportunity—typically took three to five days for each request.

Process change: Implemented formal request process that included key information from managers on what they specifically needed.

Result: Average response time reduced to less than two days, typically with higher-quality materials and improved capture rate.

Figure 6.4 Unplanned work improvement.

For other departments, a good way to start addressing unplanned work is to focus on those with the highest occurrence or highest disruption to your department. By process mapping and applying root cause analysis to the unplanned work, you can improve (implement) a process to make it unplanned.

An important element of unplanned work and department process improvement is that it never ends—there will always be unplanned work and the need to improve how the department operates. By keeping metrics on the unplanned data and focusing on the highest disrupters, you will naturally improve how you operate. An added bonus is that you will find that a portion of your unplanned work is just waste and work that should never have been done. This allows you to understand why it was not needed and clearly communicate to the requesters. Periodically review your planned work processes to ensure that they are continually improving through your daily improvement processes, as well as identifying planned work processes that are no longer needed or should go through a formal process improvement due to changes in systems, approaches, or other fundamental inflection points.

Organization

To me, organizational process improvement should be accomplished infrequently, as individual, team, and department improvement captures the majority of the opportunity for improvement within an organization. However, there are times when an organization needs to improve, and this improvement will make an impact on everyone. There are three primary focus areas for organizational improvement:

- Legacy items (see Figure 6.5)—When looking at any organization, the way you operate and communicate your procedures typically builds up over time. If you do not have good governance and oversight, these procedures become out of date, are added to because of "past issues," and eventually become large and unused.

Legacy: 30-year organization policy and procedure manual, consisting of almost 400 sections.

Challenge: Manual was outdated, with impossible to understand expectations, and just not used.

Improvement team: More than 50 core team members accomplished process mapping, policy/procedure development, and new manual adoption.

Result: Manual was reduced to 30 sections that clearly conveyed (written and graphical flowcharts) expectations of the organization on how to operate, with the ability to continually improve and incorporate best practices.

Figure 6.5 Legacy process improvement.

- Enterprise systems—While changing enterprise systems is often managed by your IT department, it cannot be treated as a departmental process improvement because the change affects everyone in the enterprise.
- Mergers and acquisitions—The hope in a merger/acquisition is that the best of both sides come together to make the whole the best it can be. Unfortunately, the reality is that one side is usually told to adopt the other side's systems and procedures, leaving both sides operating at suboptimal levels.

In all these cases, organizational improvement teams should use a cross-functional and cross-regional collaborative team to create process maps and establish flow. These process improvement teams are not just there to identify what the process improvement will be but also to act as ambassadors for actually implementing the changes and driving Quality First.

Use of Agile Teams

When accomplishing process improvement, the team members you bring together are likely doing the improvement in addition to their normal workload. This means it is very easy to get distracted and for the process improvement to take second, third, or lower priority for team members. Obviously, having full-time personnel assigned to the process improvement is ideal, but usually that is not realistic.

Therefore, in addition to the teams being collaborative, cross-functional, and cross-regional, they need to be set up and allowed to operate in an agile manner. The key attributes in setting up an agile team include:

- Flat structure—The team members set the priorities and deliverables, and they work as a team to accomplish them. This is not a top-down approach but a peer-to-peer approach—where a vice president and an administrative assistant are equal on the team.
- Process-mapping event—Start with the entire team in the same place (physical or virtual), all accomplishing process mapping together, facilitated by an impartial party.
- Scrum board—Use a scrum board to manage the work to be accomplished in two-week sprints, detailing the "to do," "doing," and "done" items—where any team member (or subcommittee member on large initiatives) can jump in and work on an activity.

- Daily (virtual) stand-ups—To advance the process improvement, do short yet consistent stand-ups (in person when possible, virtually when not) to align what people did, what they are doing, and what they need help with. Use the scrum boards during the daily stand-ups to keep everyone on the same page, as well as enabling those who could not make the stand-up to stay abreast of the changes and needs.

Using the agile team approach typically cuts the time for process improvement by 50 percent or more. Additionally, it often achieves faster and stronger adoption upon implementation, as the team continues using the scrum board and daily stand-ups through initial implementation to adjust what is working and not working. On one process improvement (Figure 6.5), past attempts took five years to accomplish organizational change, but an agile team was able to accomplish this same work in approximately nine months.

Supercharging the Transition

Using process improvement selectively can supercharge your transition to Quality First by addressing the organizational and departmental processes that do not align with your core values or your definitions of RT1. Obviously, identifying which processes to address is your greatest challenge, as it may not be apparent which processes support your values/RT1 and which do not.

Identifying the processes that you should focus on and improve early in your Quality First journey can be accomplished in several ways:

- Ask your employees: The first way, which I prefer, is to simply ask your employees. Trust me, they know which processes are not working and are not aligned with your values. You can use surveys, meetings, or small teams to obtain feedback. Your focus should be on what processes/systems are challenging to your employees and which ones do not align with your definitions of RT1 (organization level).

- Look for legacy: Another area to focus on is your legacy processes and systems. If people say, "Well, that's just how we have always done it," this would be a good starting point to identify the processes to evaluate. Legacy items can be very positive and reinforce your values, but they also can become outdated and not align with your current definitions of RT1.

- Challenge leadership: While the first two address the organization, this one is focused on your various departments, where you challenge departmental leadership to identify their processes that do not align with RT1 and need improvement. In organizations with good personal accountability and trust, this works very effectively. However, if there is a lack of trust, the leaders themselves will not identify the true processes that need to be improved.

To maintain focus and momentum on supercharging your transition to Quality First using process improvement, I recommend that you start small, with two or three process improvements; once they are completed and show progress, then work on another two or three improvements. After these are completed, you can expand to the department level and have each department do two or three. So, you should likely have 15 to 30 improvements in a typical organization as part of your initial Quality First journey.

Using Data to Celebrate Improvements

In addition to the tools previously mentioned (daily stand-ups, process mapping, one-page plans), there are helpful tools associated with data and their visualization, as well as the need and value of celebrations surrounding improvements.

Use of Data

As organizations digitize their supply chains and processes, their data are becoming hidden gold mines to better understand and drive their operations, both internally and externally. We'll dive much deeper into data in Chapter 9, "Catalyst 2: Data Intelligence," but relative to improvement, the focus of data needs to be on what to improve and how these improvements are trending.

The challenge with data is that data themselves are of little value. For example, having data on all the hours worked on a project may be of interest, but their value only comes when you can compare those hours against something that provides insight and thus value. You could compare hours with scope and other projects, or hours with profit, safety accidents, or customer satisfaction. The value of the data thus only comes from the ability to compare them with something else.

Relative to improvement, the data you want to look at typically fit into two primary buckets:

1. Trends—If you know what you are trying to improve, the data will be analyzed over time to explain the improvement made. This could be cycle times, resources required, or other key indicators. The data are analyzed to understand the direction and velocity of improvement.

2. Opportunity—Often, you may not know where to focus your improvement efforts. In these cases, you will be looking for opportunities for improvement through the data. Traditionally, you would be looking for defects, issues, and negative feedback items in the data. However, to provide focus, you need to categorize the data to know what the true issues are that are affecting your organization—that is where visualizations come in.

Visualization

While data may be the gold mine, the gold itself is found via visualization—where you can see the data in charts instead of just numbers. It is visualization that brings the data alive and enables you to find the true opportunities for improvement.

While there are many ways to visualize data, the simplest form that I use at the start of any initiative is that of a quadrant diagram, which is a simple X-Y chart comparing two variables for an item and dividing the chart into four areas (see Figure 6.6). The areas are divided by the 50th percentile of data, where 50 percent of the cumulative data are on both sides of a horizontal line, and then 50 percent of the cumulative data are on both sides of a vertical line. The four quadrants are thus:

1. Lower left (Q1)—low number (x-axis) and low value (y-axis) of items
2. Lower right (Q2)—high number, but low value of items
3. Upper left (Q3)—low number, with high value of items
4. Upper right (Q4)—high number and high value of items

You want to focus on the Q4 items (upper-right quadrant), as they have the highest impact on your organization. For example, maybe you track all warranty calls and the cost/time to resolve them. The Q4 warranty calls would be the ones to focus on to change the process to get out of your system first.

The great thing about quadrant diagrams is that as Q4s are resolved (they move down and to the left, to become a Q1), and as you reanalyze the data, new Q4s will emerge. The new Q4s typically will not have the same severity as the resolved Q4s, but the approach keeps you focused on what is hurting the organization the most.

Celebrate Improvements

For continuous improvement to flourish in your organization, the last thing you need to do is not just recognize the improvements, but celebrate them. Oddly enough, I find this the hardest thing for managers and senior leaders to do. Unfortunately, they are often more focused on negative situations and on "fixing problems" than on looking for and recognizing those doing good, and even exceptional, work. However, there are several opportunities for leadership to celebrate improvements throughout the year:

- Performance planning—One of the most effective ways to recognize improvement is to make it part of each individual's performance plan and have it as a formal discussion point: "Let's discuss the improvements you have made this year." To celebrate, the manager can identify one item from each person and convene a lunch with simple certificates in celebration of the key improvements by each individual.

Industry: The building construction industry has a peer group—the Design and Construction Excellence Exchange (DCX 2024)—that looks at improving the way construction is planned and executed.

Data: The DCX collects and analyzes member data around quality incidents—which are occurrences on projects that should not have happened if a solid quality program was followed. The incidents are tracked by scope of work and cost impact.

Q4 plot: The incidents are shown on a Q4 plot with the number of incidents on the x-axis and cost of incidents on the y-axis. Each data point is for a scope of work (CSI section).

Result: Out of $516 million in quality incidents, $146 million were Q4s—focusing the industry on 11 scopes of work versus hundreds.

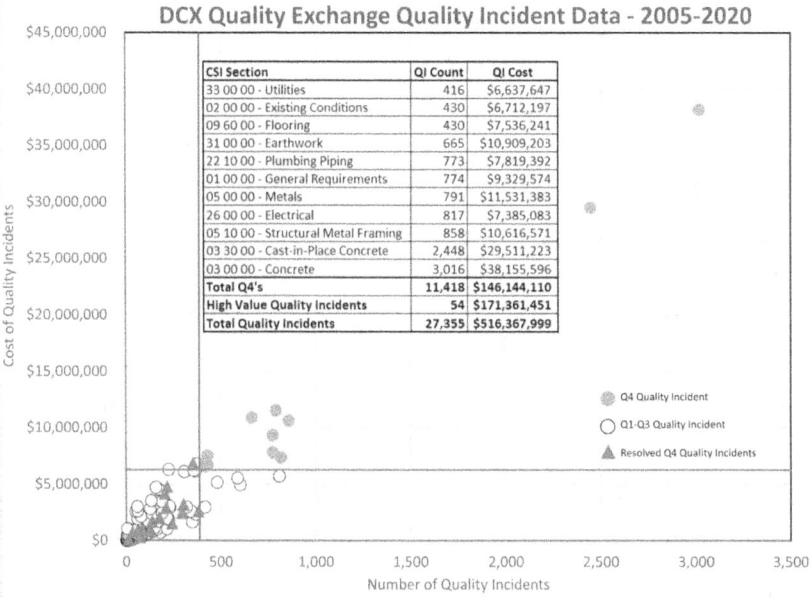

Figure 6.6 Q4 in practice.
Chart used courtesy of the Design and Construction Executives Exchange

- Newsletters/intranet—At the organizational level, it is important to have an ongoing string of recognitions. This can be through employee newsletters or on your organization's intranet homepage. These need to be visible and current. To celebrate, the best story/event should be highlighted on an annual basis, with some event where the CEO and leadership recognize those individuals/teams that contributed to the success.

- Meetings—In meetings and your daily stand-ups, simply recognize the improvements that have been made.
- Personal—As leaders see improvements, they should recognize them. This is especially important as they walk around the organization—ask "What have you improved recently?" This not only helps recognize the improvement but also sends a clear message to everyone that leaders are concerned and aware of improvements.
- Awards—At a department or organization level, you can have awards for the best improvement. This can be challenging, but if you are sincere and focused, it does send the right message.

Continuous Improvement Foundational Element Activity

It is now your turn to start your continuous improvement journey using Figure 6.7; note that this can be for an individual or a small group, but I find the small group works better, as you hold each other accountable for improving.

1. Identify the team/individuals—fill in the team name, current date, and individual names.
2. Discuss the previous day—what worked, what did not work, and what was learned?
3. Discuss today—what is each person's continuous improvement?

Remember that this daily approach to improvement allows for and expects small failures to happen and that you learn and move forward from what does not work. Responding positively to failures is essential to individuals wanting to try something new the next time.

To bring the four foundational elements together, we'll examine the performance element—everyone benefits—in the next chapter.

❶ Continuous improvement stand-up meeting	
Team name _____	
Current date _____	

Name	❷ Previous day's results
	Yesterday's results
_____	_____
_____	_____
_____	_____
_____	_____

Name	❸ To be accomplished today
	Today's focus
_____	_____
_____	_____
_____	_____
_____	_____
_____	_____

Figure 6.7 Summary of continuous improvement.

Chapter 7
Performance Element: Everyone Benefits

The *everyone benefits* performance element is a result of successfully implementing the four foundational elements, but it needs to be a focus so the value of Quality First can be known and seen. Also, a focus on how everyone benefits keeps the discussion broad and includes all stakeholders—internal to internal, and internal to external.

Importance to Quality First

How everyone benefits is essential to Quality First because it provides an ongoing reminder of why you have Quality First and the importance of maintaining a focus on driving it from your organization's core. Without a focus on the performance element—everyone benefits—leaders may take their eyes off it, especially when there are leadership changes. The culture can also quickly change, and quality can degrade and become an add-on activity, not core to the organization.

It's a Result

Primarily, the performance element is a result of the four foundational elements. By consistently implementing the foundational elements, everyone will benefit through Quality First.

- Define and Do RT1—focus on having benefit discussions, which are similar to root cause analysis focused on an issue that occurred, but it flips the discussion to "What would have been the impact if we had not achieved RT1?" These discussions not only provide a better understanding of the value of RT1 but also reinforce the value of continuing to improve your definition and doing of RT1.

- Individual/Collaborate—accomplish this through your organization's performance-planning process, as it provides the opportunity for individual discussions as well as to set individual goals for improvement. Focus on the value to the individual, so they recognize their contributions and improvements, which can then be tied to the broader team value.
- Embedded Verification—one of the best ways to identify and document the value of embedded verification is to use the NGT introduced in Chapter 3 and have an annual discussion with each team about the value they see from embedded verification.
- Continuous Improvement—the benefit of continuous improvement focuses on process improvements made in the organization by summarizing the results from the individual process improvements accomplished throughout the year.

The Wide Array of Benefits

One thing to remember and focus on as you discuss benefits is that a wide array of benefits are achieved in a Quality First organization—these go well beyond simple definitions of quality or income. Items within the wide array of benefits include:

- Monetary
- Time (personal and company)
- Employee satisfaction
- Customer satisfaction
- Supplier/vendor satisfaction
- Warranty rates
- Recall rates
- Promises kept (delivery times)
- Bottom- to top-line growth
- Market penetration
- Industry ranking
- Employee voluntary turnover
- Community ranking

Empowerment

As you look at how the foundational elements achieve everyone benefits, a common thread throughout is that Quality First is empowering everyone in the process to benefit from it. By focusing the individuals on understanding, doing, verifying, and continually improving on RT1, you are thus empowering them to not just do it but also benefit from it.

This can be a hard transition for some organizations, as empowering their employees and stakeholders also means that they need to trust them. For those who are hesitant to empower their employees, you first need to trust Quality First, primarily by consistently driving the foundational elements, especially embedded verification, and the culture and the process will self-regulate and provide the trust you need.

Maintaining a Focus

As mentioned previously, everyone benefits needs to be a focus for Quality First to thrive and for the benefit to be recognized and celebrated in your organization. This is not hard, but it does need to be accomplished.

Start with Leadership

There are countless examples of organizations that had best-in-class quality programs and were leaders in their industries, and then overnight they not only lost their leadership positions but also had substantial losses in public relations, employee retention, and revenue. In all these cases, leaders began to see quality as a line item in their organization's ledger and not coming from their core, as shown for Quality First organizations.

This is the main reason you must start with leadership's understanding and alignment to the benefits and continually focus on understanding and highlighting how everyone is benefitting in your Quality First organization. This is a win-win situation, not one where someone has to win and someone has to lose.

Storytelling

The most effective way I have found to keep Quality First in the forefront of leadership and the organization is to implement a storytelling approach to everyone benefits. While the storytelling can come from case studies or other documents created to support capturing the benefits and value, it is the storytelling by individuals that brings passion and value to the organization. Storytelling is meant to be:

- Informal—The person tells his/her story, so it should not be scripted but informal in nature, maybe with a few talking points to keep the storyteller focused.
- Personal—The storyteller needs to provide a personal perspective and why he/she sees the benefit and value. It is the personal story that people remember.
- All levels—You need to have storytelling at all levels—from the boardroom to the factory floor.
- Repeated—Stories are intended to be repeated or paraphrased by others to continue the message and even the mystique of the story.

I understand not everyone is good at telling a story; however, people do like to talk about what worked and what they did. Storytelling should be encouraged at gatherings and meetings, and leaders themselves should set the example.

Making Exceptional Fun

The last way to focus on everyone benefits is to make exceptional performance fun. What I mean by this is that you need to recognize success more than failure. You must make it clear that the organization is committed to Quality First and that you'll celebrate keeping it as part of who you are and how you operate.

To me, making exceptional fun is about small gestures at the team level that reinforce the behavior, focus, and push to do RT1. This could be in the form of team lunches out of the office or gift cards for meeting key milestones or deliverables. You could also do this at the organizational level, but this often does not generate the fun, passion, or camaraderie you are looking for.

Everyone Benefits Performance Element Activity

It's your turn again to put into practice what you've learned—use Figure 7.1 to complete an inventory of potential sources you currently have for documenting everyone benefits in your organization:

1. Organization information
2. Surveys—Identify the surveys completed by your organization, both internal and external (I or E). Document the key information obtained from each survey.

3. Industry Ratings/Evaluations—Identify each industry rating or evaluation (R or E) your organization participates in and what information is used to determine ranking (for example, revenue figures, employee surveys, and so on).
4. Business Intelligence—Identify your internal systems with data that can be accessed and analyzed for trends, both server and cloud-based (S or C). Document what trend you would be looking for.

❶ Organization information			
Name	_____	# Employees	_____
Industry	_____	# Clients/projects	_____
Revenue	_____	Average growth	_____

❷ Surveys		
Survey	Type	Key information
_____	I/E	_____
_____	I/E	_____
_____	I/E	_____

❸ Industry ratings/evaluations		
Rating/evaluation	Type	Information used
_____	R/E	_____
_____	R/E	_____
_____	R/E	_____

❹ Business intelligence		
System	Type	Data/trend
_____	S/C	_____
_____	S/C	_____
_____	S/C	_____

Figure 7.1 Summarizing how everyone benefits.

Part II
Catalysts

Chapter 8
Catalyst 1: The Power of Once

As a kid, I always loved watching science shows and seeing how the scientists would mix a bunch of chemicals and then add such a tiny amount of a substance to make it all go crazy! It was the simple catalysts that made the reaction go into hyperdrive and move at incredible speed. While science experiments and their catalysts are often regulated in a laboratory, there are catalysts in everyday life—you just need to look. For Quality First, catalysts are items that accelerate the benefits and implementation of the foundational elements by tying them together through a common purpose or idea. The first catalyst is the concept of the "power of once."

What Is Once?

To understand the power of once, we first must understand what "once" is, for though it is a very simple term, it is hard to realize in practice. When you look at all the quality programs and business improvement approaches out there, a common thread is to do work once, as I think we can all agree that rework is not something anyone wants or likes to do.

Many Different Terms

When you look across industries and organizations, there are many terms used for "once"—one is not better or worse than the other, but it is important to understand that these terms all mean the same thing—we do the work or activity only once:

- Do it right the first time
- No rework
- No shortcuts
- Build it like you would for your family
- Give them quality

I personally like the term "once" over "first time;" it is simpler and a little more direct about the expectation. In addition, I feel it has a more positive connotation, in that no one wants to take the SAT again or go back to the DMV again—they just want to do it once. On the contrary, I find that "first time" implies that there may be a second time.

It Is Hard to Do

While it is easy to *say* you should just do it once, as everyone knows, it is very hard to actually *do*. There are no perfect systems or processes, and people are inherently fallible. Therefore, completing each activity only once requires more than just a slogan, and it must be integral to how the work is planned and executed.

This is where the power of once comes in, as it overlays the foundational elements; simply implementing the foundational elements will get you closer to once. However, by consciously and verbally using the power of once, you reinforce the intent of doing all work once, and when that is not accomplished, it needs to be addressed—not in a negative way, but for learning purposes.

Accept That It Is a Never-Ending Journey

Finally, for the power of once to be fully embraced and implemented as part of Quality First, you need to accept that this is a never-ending journey. It is not something you launch and then forget about. You need to put time and effort into maintaining a focus on doing work once through your processes, procedures, and messaging.

However, this power-of-once journey is complementary to and provides a strong catalyst to your Quality First journey by providing simplicity on purpose and messaging.

What It Takes to Do It Once

To do all work once does take effort, planning, and follow-through when issues do arise. Focusing on "once" as part of Quality First starts with RT1 and then leads to ongoing conversations. Data also play an important role in achieving once because you need to know how you are performing and how close to once you are.

Start with RT1

As RT1 starts and comes from your organization's core values, to drive *once* through Quality First, you must start with RT1. The primary reason for this is that during the creation, maintenance, and improvement of RT1 throughout your organization, once shows how you will achieve RT1, which is simply doing your work once.

When you look at the definitions of RT1 for different parts of your organization, there are really two perspectives on once that you need to take:

- Do it once—The first perspective is that it is a physical activity for which the goal is to do the activity once. This would be for such things as building a widget, processing paperwork, or other procedural activities.

- Only one chance—The second perspective is when you only have one chance to do something; after that first chance, you have set in motion all future interactions. Therefore, this perspective is more for personal interactions, where if you set a non-RT1 tone, it is really hard to change from that.

A great example of RT1 and once is the onboarding of new people to your organization. If your definition of RT1 for people includes collaboration and respect for the individual, then you need to ask yourself how you will achieve these during the first interaction with the new employee. This is where the power of once shines, in that you need to address your onboarding process, specific points of interaction, and also training of those onboarding new people—all while focusing on how to use your one chance to drive RT1 for people.

This is a different perspective than is taken in most organizations today. They become very procedural in their processes and don't tie them back to both RT1 and once, which results in not conveying the organization's core values at the point in time when you really should be conveying them. Therefore, by starting with RT1, then discussing how to achieve RT1 once, you build the framework for your "how" and also begin engaging the individual and teams in focusing on improving their processes and approaches from your foundations.

Ongoing Conversations

However, because "once" is a never-ending journey, it is important to go past the definition of RT1 and have ongoing conversations about how you are achieving once. This is best accomplished at the individual/collaboration level, ideally as part of your daily stand-up.

As was discussed previously, focusing on the negative (rework) is easy for management and leadership; however, focusing and discussing the positive is often lost in the noise. Therefore, it is important to have ongoing conversations throughout your organization about what "once" looks like and how it can be achieved. I am actually surprised how many organizations do not have these simple conversations—instead, they spend weeks and months planning their work, never asking themselves if they are addressing RT1 and if what they are doing will just be once.

Obviously, they are not ignoring "once;" it is implied in what they are doing and is critical, yet it is kept under the surface. By bringing once to the surface, individuals are more likely to be open and honest about how things are actually going, and when once is not achieved, they no longer hide their mistakes but understand that paying attention to "once" is part of any process and that they can always improve what they do.

Use of Data

Similar to previous discussions of data, the use of data is essential for understanding and achieving "once" in your organization. Now, because there are so many different RT1s and onces in an organization, it is important to identify those that already have data as part of the process and can be used to support and drive once in departments and the whole organization.

The data need to already exist; or, if new, they can be integrated into current systems without additional effort to enter them. Once you add work, you are defeating the purpose of RT1 and once in many cases. By using data that are already available and part of the process, you are simply adding the analysis and the visualization of once for those in the process. Ideally, the data are presented in two formats to drive once in your organization:

- Pass rates—This has been called many things, such as first-time quality, but ultimately it is a measure of how successful you are at doing work once. This is a positive metric because it focuses on all that was done right, not necessarily the not-right side of the equation. Now it is easy to see the not-right side, but it is not the primary focus.

- Exceptions—I really like the term "exceptions," instead of wrong, failure, or another negative term. Simply put, an exception is something that happened but was not expected if the process worked. Since no process is perfect, we always expect some issues to arise. I also find that leadership is able to understand and focus on exceptions better, given that they are involved in creating what an exception is and what is expected of them when one occurs.

Addressing It Twice

There are always issues, and nothing is perfect, so you will also need to address the situation when something takes more than once to get right. Accomplishing something twice is never fun and is definitely not wanted, but this process does need to be part of your organization because it gives you a way to continually improve and focus on RT1 and once.

Do You Know Where You Are Bleeding?

When you are bleeding and go to the emergency room, it is apparent to the nurses and doctors that they need to stop your bleeding to keep you alive—it is very visual and visceral. However, when you look at your organization, do you really know where and when you are bleeding? Where are your problems, and where should you be focusing your efforts?

From the perspective of the "power of once," you need to get rid of the noise and be able to focus your efforts where the organization is bleeding the most. The Q4 charts from continuous improvement are one way to provide this understanding and focus (see Chapter 6, especially Figure 6.6). However, you may need to find other ways to see where you are bleeding:

- Ask employees—As mentioned previously, I am a big fan of just asking people. Your employees know when they are performing and not performing. Simply ask them where the issues are or where your processes are bleeding, and they will have the answer—you then just need to figure out the best way to stop the bleeding.
- Identify your roamers—You may need to look for them, but in most organizations, you have people whose job it is to roam around the company supporting your projects or initiatives. These roamers have valuable insights into what is actually happening within your organization, at either the organizational or regional level, depending on your size. You can tap your roamers to identify where you are bleeding.

- Catch-all accounts—The last place to look for bleeding is in your accounting system. The first place to look is at any catch-all accounts, where people put their time or expenses when there is no other place to put them. These could be closed jobs, warranties, or miscellaneous categories. By taking a deeper look at what specifically is put into these accounts, you can often find bleeding within your organization.

Once you identify where you are bleeding, you can then use the foundational elements and Quality First to address and stop the bleeding.

It's How We Have Always Done It

There is a special category that needs to be addressed when looking at the "power of once." When you look, the work is being accomplished at a high rate of "once," yet there is a huge opportunity you are not tapping into. Call it complacency, comfort zone, or something else, but it is when anyone in your organization says, "Well, that is just how we have always done it."

In these cases, everything looks good, but not great, because your process is likely stagnant, and you are not implementing continuous improvement on how you are doing the work or defining RT1. Therefore, it is important to understand the opportunity being missed and to challenge individual staff members and teams to fully embrace improving RT1 and how they accomplish their work. Start with daily stand-ups, and go from there.

The Small Wins Are Huge

The "power of once" really comes from the small wins. Through continuous improvement and daily stand-ups, you are focusing on RT1 and once with every individual in your organization. The cumulation of all these small wins is huge for your organization.

Think about your work and any instance of having to do rework. What was the impact on you personally, the project, and your organization? Rework often constitutes two to three times the effort to do it once, and you never have enough time to do it once, much less twice. Maybe you have five of these in a typical week—what is the value to you if you could reduce that to four?

This is the power of once at its heart—the individual focuses on doing a task only once and on continually improving how he/she can do the work to achieve that.

The True Value of Quality First

In looking at Quality First, the "power of once" is where the true value is generated. Through defining and doing RT1 by the individual and continually improving upon it, with a constant focus on once, you are creating the most effective approaches to your work that are possible. Nothing is perfect, but the Quality First foundation will get your organization closest to perfection.

Once Is Where Value Is Lost

A fundamental tenet of the power of once is that anytime you have to do something more than once, you lose value. This lost value comes in many forms:

- Time
- Materials
- Reputation
- Clients
- People (leave)
- Capital
- Profit

The power of once focuses on how to achieve all work both RT1 and once yet must include learning from those times when work took more than once to accomplish. This learning flows back into Quality First to continually improve its focus and implementation.

The Power of Once Transcends All Elements

Because the power of once overlays the foundational and performance elements of Quality First, it also transcends them as a binder and a focus to understanding and reinforcing the elements, in these ways:

- Right—We've reviewed extensively how RT1 and once are intertwined. Once provides the point of discussion on how you and your teams will achieve RT1 and address any issues when RT1 is not achieved.

- Individual/Collaboration—Ultimately, it is the individual who determines whether his/her work is done once or requires more times to accomplish. By focusing on the individual's understanding of RT1 and the power of once, you enable the individual to continually strive to improve and achieve RT1 once.

- Embedded Verification—By embedding verification into your systems and processes, you are enabling a mindset and achievement of once. The key is that you have to keep the focus and communication on the positive side of the achievement of once, not the failure of more than once. In addition, you keep leaders focused on exceptions where they can make a difference.

- Continuous Improvement—As we recognize that processes and people are not perfect, continuous improvement with failures to achieve once provide the focus needed within Quality First and what to improve. Ultimately, when you look at continuous improvement and once, you get a clear understanding of the "Why" for continuous improvement through any failure in once. This focus and understanding are key to keeping improvement continuous.
- Everyone Benefits—The power of once extends well beyond just your employees. Think about individuals calling your organization with an issue. Your goal should be that they only need to call once to resolve their issue or answer their question. Think about how many times you have had to call to get an issue resolved. With the power of once, maybe you change your process; if you cannot resolve their issue immediately, you take responsibility to figure it out and then call them back. They still only call once!

Chapter 9
Catalyst 2: Data Intelligence

There is a huge and real challenge of data: We actually have too much data, in too many systems, and we cannot really get value from the data we have. This is where data intelligence catalyst for Quality First comes in.

Data intelligence is using your data to understand your trends, identify where you need to focus your efforts, and ultimately drive the foundational elements with visualizations and messaging from the data themselves.

Here, I need to make a distinction between the field of business intelligence and what is needed for Quality First. For the most part, business intelligence is focused on how to drive the business, create a better understanding of it, and ultimately generate more revenue—just look at how the technology giants almost solely rely on business intelligence and artificial intelligence to drive their platforms and the user experience. Data intelligence is a subset of business intelligence but with a focus on your foundational elements, and it is used to enhance how you drive them from your core. Therefore, though you may use the same data as others, your focus is different because it includes both inward-looking and outward-looking data.

Identifying Metrics

The biggest challenge for data intelligence is identifying the best metrics to use. Often, this requires some level of trial and error, but there is a road map you can follow to help keep you on a narrower path in finding the right data and metrics. The data intelligence road map includes these key steps:

- Create a Quality First data map
- Prioritize the data

- Ensure data reliability
- Do metric investigation
- Implement metrics

A Quality First Data Map

The one thing that amazes me the most about data is that when I talk with companies about their data, they don't really know what they have. They know they have systems, and they know they have a bunch of data—but typically, beyond financials, they don't fully understand their data.

Therefore, the first step is to create a Quality First data map, which is a subset of all the data available and which focuses on the relevant data that are aligned with your foundational elements. A good way to start creating your Quality First data map is to identify all your current systems/processes that produce/manage data. Once the high-level systems/processes are identified, then you can summarize the key data that are within each.

This exercise provides you with a high-level understanding of your data ecosystem, which you can then use to identify the key data relevant to your foundational elements. As a starting point, you can create a simple matrix, which you can then expand on as you go through this process (Table 9.1).

Obviously, in larger and more complex organizations, this "simple" matrix can get quite complicated. Keep it simple and at the highest level possible to start with, and then break it down as you dive into each definition of RT1 within each department. For a first pass, keeping it at the organizational level will provide you with the primary items to focus on.

It is important to note that the Quality First data map is about the data already being created and managed within your organization—it is not about adding new processes and efforts in capturing/creating new data. While there may be cases where you are required to identify new data, it should be integrated into an existing system/process so additional effort is not required. This ensures greater reliability and completeness of the data.

One final thought about the Quality First data map—as you brainstorm each foundational element, don't just focus on what you have as raw data, but think about what additional data (and thus metrics) you may need to better understand and drive your foundational elements. Some potential examples include:

- RT1—You likely have communications of what RT1 is, but are you categorizing (tagging taxonomy) these to capture relative content created, delivered, and received?
- Individual—You may have employee start dates, but you may also need data on tenure with the organization and tenure in position to better understand performance.

System/ process data	Foundational element			
	Right	Individual/ collaboration	Embedded verification	Continuous improvement
System 1				
Process 1				
Data 1				
Data 2				
Process 2				
Data 1				
Data 2				
System 2				
Process 1				
Data 1				
Data 2				
Process 2				
Data 1				
Data 2				

Table 9.1 Matrix of your data ecosystem.

- Embedded Verification—Most organizations capture the results of a verification (pass/fail rates), but you may need to calculate and focus on pass rates (first-time quality) more than on failures.
- Continuous Improvement—A key part of continuous improvement is sharing knowledge and lessons learned; do you need data on how many knowledge articles individuals post and how many they consume?

Prioritize Data

Once you have your Quality First data map, the next step is to prioritize the data, given that most of the time you will have identified more data than you can actually analyze. Unfortunately, this is where the iterative nature of data intelligence comes in, as you'll need to start with your high-priority data, go through the remaining steps, and then come back and sometimes start over or at least tweak your approach with other data until you get to your core metrics.

Prioritizing your data is subjective, and you need to use your best interpretation of how the data can help you drive your foundational elements and Quality First. If you have already gone through the process of understanding and starting to implement your foundational elements, you will actually be adept at prioritizing the data, as you will quickly see if they are relevant based on your definitions and implementation. Put another way, you cannot start your Quality First journey with data; instead, you must go through all the steps identified in this book to make data intelligence a catalyst.

In prioritizing your data, I recommend a simple three-level scale—low, medium, and high. I find that trying to use more complex ratings (for example, 1–10) not only makes the prioritization process more complicated but also often results in less clarity and focus. By simply defining the three levels in this way, you can quickly prioritize your data:

- Low—the data have low relevance to the foundational element or are of little value in driving the foundational element
- High—the data have high relevance to the foundational element and are likely to have high value in driving the foundational element
- Medium—all other data

The medium level is the main reason I like three levels—you just need to identify two levels (low and high), and the rest go in the middle. In addition, when you do the prioritization, you are focusing on identifying if particular data are low or high priority. If they are, then great; if not, then they are medium. This reduces a lot of back-and-forth discussion and gets you to the results more quickly.

Once you have completed the initial prioritization and have low/medium/high for each data row and each foundational element column, your next step is to review the entire data map to identify the overall high-priority data. These overall high-priority data are those data rows that have multiple highs (that is, more than one foundational element), which indicate that by using these data, you are crossing over the foundational elements and are providing higher-value indicators. Again, this is subjective, so after you look at your data map, you will typically prioritize those with either two highs or one high and two mediums. As a best practice, I use color coding of the data map cells to visually see this on the spreadsheet. For example, I use a green highlight for high, a yellow highlight for medium, and clear for low.

While there is no right or wrong number of overall high-priority data items, I do find that once you are above 20 to 30 items, you are likely diluting what can be understood and focused on. In smaller organizations, 10 is probably a reasonable number.

Data Reliability and Completeness

Before expending efforts to dive into the data and investigate metrics, the next step is to pause and discuss the reliability and completeness of the data:

- Data reliability—When looking at the reliability of the data, you need to ask yourself if you can trust the data, whether they are accurate, and whether the processes generating the data are repeatable.

- Data completeness—When looking at the completeness of the data, you need to ask how complete the data set is. Are you capturing all instances of occurrences or are they sparse?

When you look at the combination of reliability and completeness, you can come to understand the usability of the data for your purposes. Again, there is no right or wrong answer to these questions, but what is important is to have a discussion. I have found cases where there is high data reliability but low completeness (20 percent), but the usability of the data is high, given that the 20 percent of the data I get provides me with a good picture of the whole (I trust the data I get to represent the whole). However, some cases have high reliability and completeness; however, when looking at the data, their usability is low (likely, most of your low-priority data).

By the end of your usability exercise, you should have a good set of data with which to advance to the next step.

Metric Investigation

Once you have your short list of data, the next step in developing your data intelligence road map is metric investigation. There are some slight semantics here. I see the data as raw information, whereas metrics entail analyzing and understanding the data. This could be a change of value over time, comparing two data points (for example, events and cost), or transforming the data (for example, linear to logarithmic).

Thus, a metric investigation is playing with the data to see what you can understand from them. The tools you have at your disposal are limitless, but I recommend you start with these:

- Time data plots: Simply plot the data over time and look for trends in the data. For instance, are they seasonal, are they trending up/down, and so on.

- X–Y data plots: Compare two data points with each other to identify relationships between them. You can also do a 3-D (XYZ) chart, but that often complicates the relationships and makes it harder to discern trends.

- Histograms/Pareto charts: Histograms and Pareto charts are great ways to see what parts of the data are important or driving the overall metrics.
- Q4 charts: Based on what you see in the X–Y plots, you can divide these into four quadrants and create Q4 charts to focus on the high-priority items within the metric.

The intent of metric investigation is to determine how you will analyze and use the data and present them to your organization for understanding and to drive your foundational elements. In addition, through this process, you will likely identify ways to improve data collection and usability.

Metric Implementation

The last step in your data intelligence road map is to implement the high-value metrics you have identified. This will include introducing them to your organization and then using them throughout the year to update the organization on progress and drive Quality First.

One of the biggest challenges in metric implementation is maintaining the value of the metric and keeping the information fresh/current. As with anything, if the metric is not adding value to your organization and it takes effort to create, it will not be used or help. Therefore, part of metric implementation is periodically reviewing the overall data intelligence map and reevaluating which metrics to use. This is especially true when major changes are made to systems/processes where the fundamental data being captured/managed change.

Because it is a fairly large effort to review and get your data intelligence map updated, I recommend that you make this an annual or biannual effort. In some cases, the span between updates may be as long as five years, aligning with and supporting an organization's strategic planning process. What is essential is that the review and update are accomplished, not necessarily the time frame—though I personally would not go past five years between efforts.

Analyzing Metric Trends

Using the same approach as in analyzing metrics, it is important to review the trends of your key metrics over time. The frequency will depend on the pace of change and the impact the metric has on your business. In manufacturing, through the use of control charts, the frequency is almost continuous; in other applications, it may be a monthly, quarterly, or annual review.

Obviously, you should have an expectation of which direction the trends should be going (up, down, or steady) and address any abnormal change in the trends. However, an essential aspect of analyzing trends is to have a discussion about the value of change. This will bring the everyone benefits performance element into the data catalyst.

There are two ways to look at the value of change—direct and indirect. The direct value of the change is a clear calculation of value generated by the change, which could be increased business/profit, improved employee retention, and so on. The direct value typically comes from known values, is often easily understood by those who see it, and does not need justification beyond the value itself.

Indirect value is a bit harder to calculate and understand because it is often focused on calculating the value of something that did not happen because you put Quality First in place. In addition, indirect values include non-direct costs such as people's time and productivity—items that are built into your system and are often hard to break apart. A classic example of indirect value is embedded verification, where an individual avoided doing rework by understanding and doing RT1 through embedded verification. If the individual did not do the rework, then there was no cost or effort for rework. How do we calculate that?

While it is often an estimate or even a guess, there is extreme value in calculating indirect values, because that is where your greatest progress is made—those items that did not occur. It is also your greatest ally in continuing to drive Quality First—the reason not to slip back. The best guidance that I can give on indirect values is to keep them simple and in an order of magnitude. For example, was it $100, $1,000, or $10,000? I find that trying to get to exact numbers takes a lot of time but doesn't provide better results at the overall level. A good way to think about indirect values is that we are willing to compromise data reliability (exactness) to make them more complete and thus more usable in the end.

The Broader Data Lake

As your organization grows and you expand the number of systems, it gets harder to understand and use the data available to you. This is when you start looking at the business intelligence tools out there and creating your own "data lake" to better understand, manage, and get value from all your systems and data.

A *data lake* is simply a virtual consolidation of the relevant data from each of your enterprise systems, spanning from email, to social, structured data, servers, and even machine/internet of things (IoT) data. It is hard to access, compare, and analyze data across different systems given that they all use proprietary data structures and approaches. By consolidating the data into a data lake, you are able to quickly access, combine, and analyze all the data, regardless of the source.

For security and other reasons, the information in the data lake is typically a replica of what is in each system and is synced either daily or more frequently, depending on business intelligence needs. Using business intelligence tools, you are able to clean, format, and cross-reference the data so they are usable and comparable across different systems. This requires you to identify keys that connect the data across your disparate systems—such as part number and employee ID—and this is the largest effort in creating and maintaining your data lake.

Regardless of your organization's size, you should create and maintain your data lake, for it provides a gold mine of information over time on how you are performing and allows you to continually evaluate the metrics you use to improve your understanding and drive Quality First. Especially for small companies, setting up your data lake early on enables you to think more strategically about the creation and use of data as you grow and add systems.

Communication

The final part of your data catalyst is communicating with your organization using all that you have implemented vis-à-vis data. For me, less is more, as it helps provide quick understanding and direction. What I mean by this is: Do not present a bunch of numbers (I know accountants and engineers love their numbers; I am one of you), but present graphically to represent action to be taken or value achieved.

Yes, this is hard; but having gone through the complex process of identifying your key metrics, creating a data lake, and using the power of business intelligence tools to represent your metrics, you will have everything you need to communicate with your organization. As shown in Figure 9.1, a great example is using Q4 charts over several years to show how your Q4 risk (those in the upper right) becomes lower risk (move to the lower left). Showing this graphically communicates visually to quickly engage your organization in realizing the value Quality First has brought to the organization and each individual's efforts.

When viewing the Design and Construction Excellence Exchange's data over time (from Figure 6.6), you can see key trends on the change in the Q4s with programs put into place by the member organizations. Figure 9.1 shows this change over time from 2010–12 to that of 2018–20 (DCX 2024).

The result is a significant decrease in risk to member organizations and the realization that the organization that identified the Q4 first was two years ahead of other member organizations and four years ahead of insurance claims.

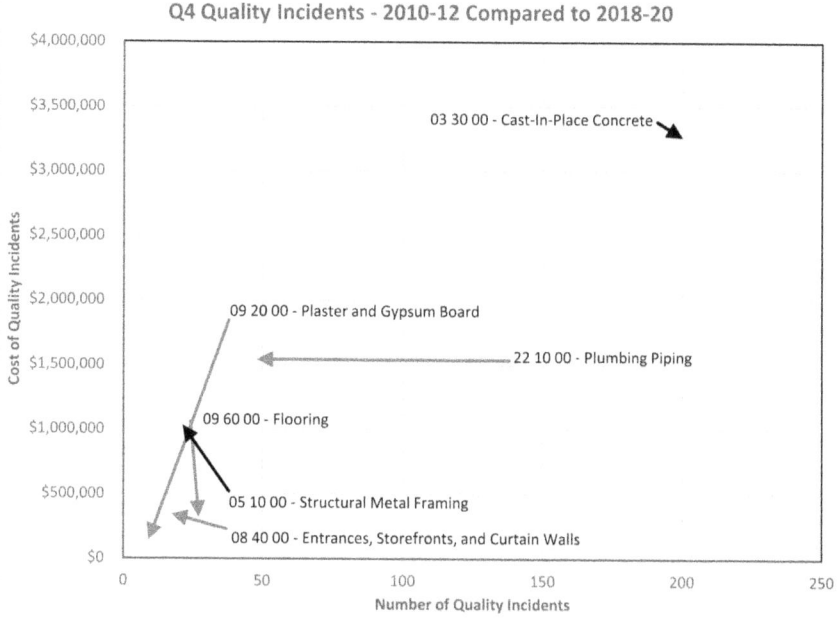

Figure 9.1 Q4 trends.
Chart used courtesy of the Design and Construction Excellence Exchange.

Chapter 10
Catalyst 3: Knowledge Snippets

The final Quality First catalyst is "knowledge snippets," which may be a new term for many of you. Simply put, a knowledge snippet is some form of information that can be provided to an individual as he/she performs a scope-of-work (preferably immediately before doing the work), to better understand how to do that work RT1 and "once."

Obviously, what a knowledge snippet actually is and how it is delivered to the individual will vary with each organization, and likely each department within the organization. For knowledge snippets to be useful, you need to focus on knowledge at the point of need, creating a knowledge organization, and integrating your knowledge snippets into your systems.

Knowledge at the Point of Need

As introduced as part of the individual foundational element, the knowledge snippet needs to be provided as part of the knowledge nexus at the point of need, which is easy to understand but is one of the hardest things to accomplish in any organization. It is easy to understand because it is simply the knowledge someone needs to do his/her job. This could be information, a procedure or process, or a technical skill. For example, think about the last time you had something new to do—say, a new expense report system. To properly complete your expense report, you would likely need to know:

- How to log in
- Steps in the process

- How to find cost codes/allocations
- How to document receipts
- Which approvals are required
- How to submit

If you are a manager, you would also need to know how to approve/reject expense reports.

However, there is a difference between having knowledge and implementing a knowledge snippet. Having knowledge is simply knowing the information is available, but implementing a knowledge snippet provides that information at the specific point of need. In the expense report example, in most organizations, you would be given training and then be told to do your reports—you are given the information, but typically not when you need it.

To get a knowledge snippet and information at the point of need, you will have to rethink how that knowledge is created and provided to the individuals "at the time and location" when they need it. Some examples of knowledge snippets for the expense report could be:

- Graphical process flow in the system—By having a simple step-by-step flowchart integrated into your expense report system, you provide the information needed to complete the process every time an individual uses the system, which is especially important as the process changes/improves.

- Pop-up knowledge boxes—In the system, you can have information (question mark) boxes pop up when individuals need additional information about what an item is or how to do a step. This could also be segments of the training/videos to refresh their understanding of the system.

- Verification screens—These can be used as part of embedded verification, in that they help the individuals verify that they are doing RT1 as they complete the steps.

As you can imagine, getting the correct information at the point of need is extremely difficult. Even with repeatable processes, such as expense reports, this can be challenging. However, it is especially challenging when the type and complexity of work are continually changing, such as in the construction business, with the primary challenge being that the individual does not know what knowledge he/she requires at the point of need.

Therefore, in instances where individuals do not know what they don't know, my recommendation is that you integrate knowledge creation and use it in the planning process for the particular scope of work:

- Current knowledge—Typically in an organization, there will be knowledge that has been created but may not be available. I always like to start with what the organization knows—find the internal resident expert and simply ask, "What are 10 things I should focus on for this scope of work?" The idea is not to make experts out of everyone but to have everyone benefit from the expert's knowledge.

- Planning knowledge gap—During the planning phase of the work, individuals should be able to identify their knowledge gaps. This is often a team effort, for you are looking at collective knowledge, and the sharing of this knowledge, to gain an understanding of the what and the how for the scope of work. Any gaps need to be identified and filled by the end of the planning phase.

- First in place/mock-up work—At the point you do the first example for the scope of work, either in place or as a mock-up, it is important to apply the knowledge gained at the point of need to verify that it meets your need of achieving RT1—if not, additional effort/learning is required.

- Share back—For others to use the knowledge you created/used, you need to share this with others, hopefully in a format that works for both you and them.

A Knowledge Organization

Keeping a steady stream of knowledge at the point of need requires that you create a knowledge organization—one that seamlessly creates, uses, and updates knowledge as a core element of the work. Creating and sustaining a knowledge organization requires focus and effort, and it is extremely difficult to accomplish, much less maintain. However, the value proposition is incredible: for everyone to have the knowledge they need, when they need it, to do their work right.

Although you may think that building a knowledge organization requires significant investment in tools and software, that is actually the easy part of the equation—the harder part is the people side and enabling a culture of knowledge in the organization:

- Culture—As with all Quality First items, to be successful, knowledge needs to be baked into an organization's culture. Therefore, knowledge creation and use need to be an element within and from the culture in order to thrive in the organization. Depending on the organization, knowledge could be a core value by itself or a sub-bullet to a core value.

- Expectation—With the culture providing the foundation, leadership still needs to set the expectation of the employees with respect to knowledge. At a minimum, the expectation should be that every individual will use and add to the knowledge of the company—and I would add "to achieve RT1" to tie it back to Quality First.
- Performance—To provide support and clarity to the expectation, it is best to integrate the use and addition of knowledge as part of individual and departmental performance plans. This provides a clear understanding to all of the importance of knowledge and provides a means to have ongoing discussions on how individuals are using and creating knowledge.
- Tools—Until you get the culture, expectations, and performance aligned and working, you should not put too much effort into tools. My recommendation is to keep tools simple until you get the culture moving so you can adopt the best tools and systems based on the culture that develops within your organization.

Ultimately, to be a knowledge organization, knowledge must continually flow through the organization. The challenge is to balance the pure collection of information with the true use of the knowledge required. To aid in finding this balance, periodically evaluate knowledge at the individual, department, and company levels using a stoplight approach:

- Individual—Simply ask individuals to rate their use and creation of knowledge in their daily work with the stoplight being the rating system: green, yellow, and red. While subjective, green means they feel really good; red, really bad; and yellow, in between. Then ask why.
- Department—I recommend that department leaders periodically look at the creation of and change in knowledge over time to evaluate how well their department is performing. Red is very little change, green is substantial and substantive change, and yellow is in between.
- Organization—At the organizational level, the focus should be on the cultural side and how pervasive the discussion is about the use and creation of knowledge. If it is a daily thing, it is green; if it is rarely talked about, it is red.

Becoming a knowledge organization is aspirational, for you will always be striving to become better at it and find better ways to gain knowledge at the point of need. There is no perfect way to do this—just continually try to drive the culture and improve the results. For me, simplicity is the best approach, yet simple is not easy.

Integrating Knowledge into Systems

Finally, you will need to integrate the knowledge snippets into your systems. I have found that adding information and guidance to systems is quite easy, but finding knowledge that is changing and improving is difficult. These are the key challenges:

- Gatekeepers—The biggest challenge with knowledge is putting in gatekeepers to "check and validate" the knowledge. Whenever you have gatekeepers, you slow down the creation and change of knowledge; but more importantly, you instill a culture of doubt into your employees because you don't trust them to add the correct knowledge. I find that eliminating gatekeepers and going with a "wiki-style" approach where the group maintains the knowledge to be accepted is much quicker and provides longer-term benefits.

- Good enough—Another challenge is that it is hard to create knowledge to start with, much less continually improve upon it. This often results in people saying what they have is good enough and not putting in any more effort. Simplifying the entry method for knowledge does not fix this, as there is often an underlying issue of not having enough time or having too many other priorities to put in the time. Therefore, not accepting the idea of good enough and driving for improvement are essential.

- Another system—Whenever you need to go to another system that is outside your normal work process, you often will not do it. Or, even worse, you use the system with which you are most comfortable (for example, a web browser) to find the information, which could be correct but is not your organization's definition of RT1. Therefore, integrating the required knowledge into your enterprise systems will often provide the best results.

There are several different ways to approach knowledge and your systems, ranging on a scale from segmented to integrated, as shown in Figure 10.1:

- *Individual drives:* Every individual typically has someplace to keep files or information—in the "old days," this was file cabinets; now it's smartphones and computers. At the far end of the segmented side, individual drives are only accessed by others by asking the individual with the information.

- *Mentorship:* The transfer of knowledge from person to person using mentorship can be a very effective means for your organization. This can be informal, where individuals connect when they want, or more formalized, through a system to match mentors and mentees. Either way, mentorship is on the segmented side of knowledge systems, because the knowledge is rarely captured to be shared and used beyond the mentee.

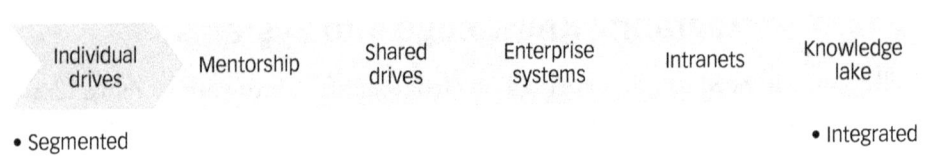

Figure 10.1 The range from segmented to integrated knowledge.

- *Shared drives:* Organizations have shared drives to keep a group's files and documents in a single location that each team member can access. This starts to move us toward the integrated side of the spectrum because you don't need to ask someone to find the information. However, shared drives often have problems with the ability to find the information, as well as updating the information as it changes. Versioning control and having clarity on nomenclature helps, but it will always be a challenge.
- *Enterprise systems:* Integrating knowledge into enterprise systems, as in the example of expense reports, is a very good way to drive knowledge to the point of need. The challenge is determining how to integrate the knowledge and then continually updating it. As these are enterprise systems, there are often strict controls on who can change what that can bog down efforts to improve knowledge and sharing.
- *Intranets:* A specific type of enterprise system, these can actually be more integrated on the spectrum, as they often enable greater flexibility in structure and maintenance of information by the masses. This can be through actual knowledge systems or somewhat ad hoc wikis and blogs. To truly keep intranets on the integrated side, they must be structured and implemented to be consistently used by all in their day-to-day work lives.
- *Knowledge lake:* As you may have realized by now, knowledge can be quite amorphous and hard to get a handle on, much less deliver at the point of need. This is very similar to the role of data in your organization (a subset of knowledge, you could say). As was introduced when discussing the catalyst of data, the creation and use of data lakes and artificial intelligence is making access and understanding cross-functional data much easier. A similar concept can be applied to knowledge—creating a knowledge lake that consolidates your key knowledge across systems (share drives, enterprise systems, and intranets) and presents them when and where they need to be—at the point of need. While there are many documents and knowledge management systems currently on the market, knowledge lakes are in their infancy and will become more robust as technology, AI, and knowledge processes improve.

The key to integrating knowledge into your systems is to first understand what your systems are and what knowledge you have in them. Similar to a data map, you can create a knowledge map to gain this understanding. From the knowledge map, you can then begin to address where you are on the integration spectrum and create your road map to further integration and the use of knowledge in your organization.

Part III
Your Quality First Journey

Chapter 11
A Road Map to Quality First

Congratulations are in order—you have made it to the point where you truly begin your Quality First journey and start to supercharge your organization from its core. You should have a good handle on the foundational and performance elements, the catalysts, and how they relate to you and your organization. Now you need to create your personalized road map to Quality First for your organization.

Before I get into what the road map is and how you do it, it's good to have a quick discussion on keeping it simple. Hopefully, by now you see that Quality First is not complicated. You have four foundational elements you focus on, and it all comes from your core. Therefore, the implementation of Quality First should remain simple. Put another way, the more elaborate and complicated you make it, the less likely it will stick and truly come from your core.

The simple elements of creating your Quality First road map (Figure 11.1) consist of starting with leadership, making your foundational elements personal through your core, prioritizing what to focus on, establishing a 1-5-10 plan, identifying the best tools to use throughout implementation, and establishing Quality First leadership.

Starting with Leadership

All Quality First organizations start with "leadership"—it is in quotation marks here, because we all know that leadership means others follow you from the example you set and the vision you have, and it's not just because of the title or position you hold. I have seen greater leadership from the lowest person in an organization than from the CEO. Therefore, Quality First does and must start with "leadership."

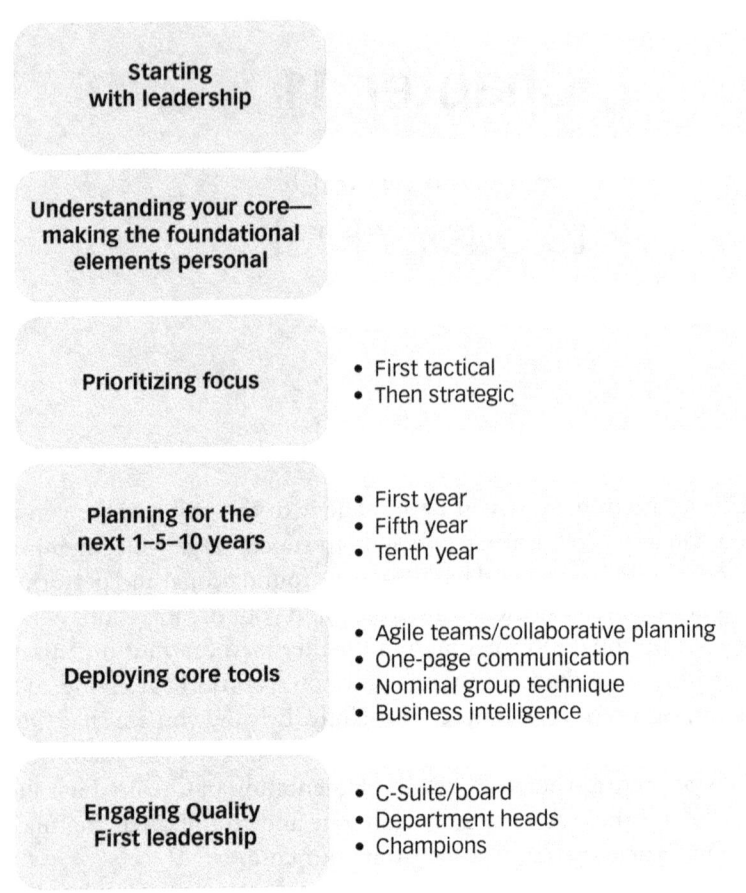

Figure 11.1 The simple elements of a road map to Quality First.

This means you need to identify and fully support your Quality First leader—the person who will facilitate your organization through the transition and enable you to become a Quality First organization. Ideally, this will be an existing and respected senior leader, because he/she brings immediate credibility to the need and change he/she will be leading.

The one thing I always think of when discussing the Quality First leader is the Broadway play *Hamilton*. In the play, Aaron Burr sings about the time Hamilton, Jefferson, and Madison made a deal on the financial structure for the fledgling United States (Wall Street in New York, for Hamilton) for the location of the nation's capital (Washington DC, near Jefferson's and Madison's Virginia). The song is about wanting to be "in the room where it happens." Now, for a cultural change such as Quality First, its leader does need to be in the room where it happens, as he/she needs to understand what is happening in the organization and be able to influence the discussion and the decisions that are made.

Therefore, at a minimum, the Quality First leader will need to be engaged, if not part of, your organization's board and senior leadership to gain alignment with the road map and to have ongoing discussions on progress, challenges, and direction. It is through these discussions that you start and grow from your core and maintain a consistent terminology using the foundational elements and the definition and doing of RT1.

Once your Quality First leader is identified, his/her first activity is to establish the core leadership team for Quality First. While you will want understanding and concurrence from the board, your real leadership team will be the C-suite, operational leaders, and department heads—it really depends on what your organization calls each of these and how large that group is, but they will be the ones you must engage and will ultimately drive the change required to become Quality First.

The best approach I have seen for the leadership team is for the Quality First leader to be the facilitator of the change, with the overall leadership team being responsible for its implementation and success. While the Quality First leader should be (and really needs to be) a peer within the leadership group, for the purposes of the planning and transition to Quality First, it would be ideal to make this his/her full-time role for two to three years during the transition. There are two ways to approach the Quality First leader:

- Moving up—One of the best choices to be the Quality First leader is someone moving up in the organization but not yet at the top level. Giving this high-potential leader the Quality First transition is both a challenge and an opportunity—the challenge being the hard work and a lot of resistance to change he/she must overcome, and the opportunity being that he/she touches every leader and corner of the organization through the change, so he/she will know the organization better than anyone else at the back end of the transition.

- Moving out—Though not as good, the other option is to tap a senior leader who will be moving out in several years, typically because of retirement. The benefit of using this person is that he/she should have credibility within the organization. The challenge is that you need to make sure this person has the skill and passion to drive the organization through the change. The downside of using the moving out versus the moving up person is that in the end, all the knowledge and connections he/she makes are lost when that person leaves the organization.

Understanding Your Core: Making the Foundational Elements Personal

With the Quality First leader and leadership team established, his/her first activity should be to ensure the team understands the organization's core and then make the foundational elements personal to the leadership team's members and to the organization. This understanding is best accomplished by using the nominal group technique (NGT) introduced in Chapter 3, with the Quality First leader being the workshop's lead facilitator and identifying a documenting facilitator to support them. The key sections of the workshop would include:

- Core values and RT1: Using the approach from the RT1 foundational element discussed in Chapter 3, the group would brainstorm their core values, identify their RT1 categories, and create their RT1 statements. This provides the main structure for their Quality First journey and often takes one to three days, depending on their current state and how well their core values accurately represent the organization.

- Defining and recognizing the individual: Getting a handle on the individual foundational element requires the team to go through a few questions that will help them understand their approach and understand the "individual" within the organization. A few NGT workshop questions that could be used include:
    – How do we currently empower individuals to be successful?
    – What are some opportunities to empower individuals to be successful?

- Embedded verification: Similar to the individual, the team needs to understand how the organization will approach and implement embedded verification so it does not become a check-the-box mentality or overly prescriptive (that is, like an inspection). NGT workshop questions that could be used include:
    – What are the easiest ways to embed verification in the organization?
    – What are the obstacles to embedding verification in the organization?
    – What activities can be eliminated through embedded verification?

- Continuous improvement: How you address the continuous improvement foundational element will depend on how continuous improvement is seen and accomplished in the organization. In organizations with a robust and active continuous improvement culture, there may be very little discussion, whereas in others, you may need much more discussion than is covered here. However, most organizations will get value from going through the following NGT workshop questions:

- What are examples of continuous improvement in the organization?
- How can we get continuous improvement to be addressed every day by each individual and team?
- How do we tell our continuous improvement story?

• Everyone benefits: As discussed in Chapter 7, even though the everyone benefits performance element is a result of the foundational elements, it still needs to be discussed to keep the value proposition understood and advancing. The NGT workshop questions that could be used include:

- What is the value proposition of Quality First for our organization?
- How do we capture and communicate that everyone benefits?

At the completion of these workshops, the Quality First leader is responsible for summarizing the results and creating the organization's Quality First guiding document, which provides a high-level view of how the leadership team defines the organization's core values and foundational elements. Ideally, this should not be more than six pages long, to keep the message succinct and easily understood by all. A check-in and concurrence by your board will be essential at this point to ensure alignment, understanding, and commitment, especially if core values are being updated.

Because Quality First comes from your organization's core, this guiding document typically does not change much over time, as it is at a high level and is based on your core values. The primary value of the document is that it provides the framework in which Quality First will be implemented, simplifying terminology and providing a single point of reference on the what and why of Quality First.

Prioritizing Your Focus

To get as much traction as possible as fast as possible, it is important to prioritize your focus on the implementation of Quality First. Unfortunately, there is no single answer to this, because every organization will be at a different place vis-à-vis each foundational element at the start. However, a general approach that I find works well is to start first with a tactical approach and then go strategic.

First, a Tactical Approach

I'm sure many of you are thinking this sounds backward—with all the focus on core values and foundational elements, why not start with strategic? Well, the simple answer is that strategy takes a long time to get right and to change, whereas a tactical approach can get you quick wins while working through the strategic changes.

To understand where to prioritize your focus, you need to understand where your challenges are and where each foundational element is on its spectrum. For example, if your organization has a robust continuous improvement culture, I would not start there. While very subjective, to understand priority opportunities, I would have the Quality First team create a simple matrix that shows each core value and RT1 statement, comparing current implementation opportunity potential (a high value means you have not addressed or implemented the item) on the horizontal axis and opportunity value (a high value indicates implementing it would add great value to the organization) on the vertical axis, as shown in Figure 11.2.

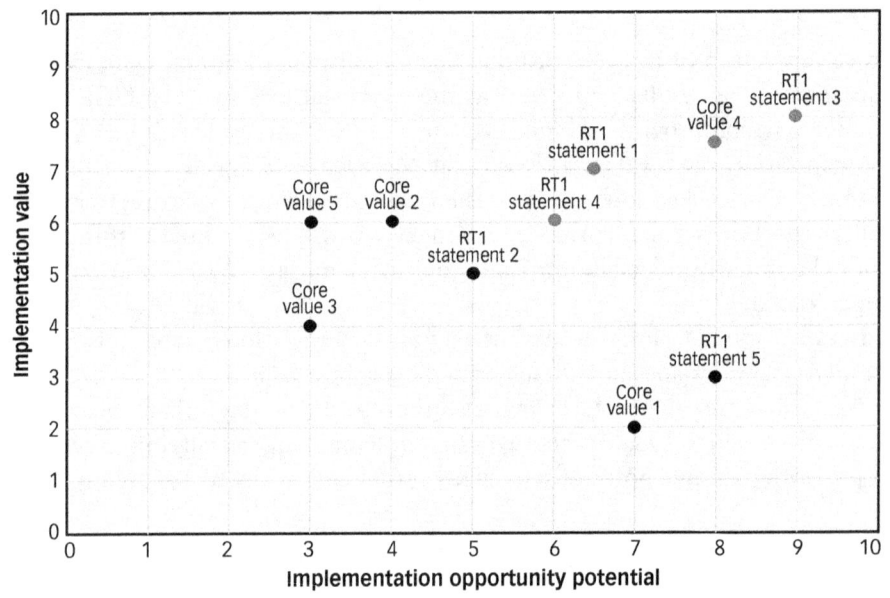

Figure 11.2 Identifying tactical opportunities.

For any item in the upper-right-hand corner of Figure 11.2, you need to clearly define its current state and what the opportunities are (typically more than one opportunity for each RT1 statement or core value). This provides a clear understanding of the item and allows for discussion of what to prioritize. Again, as this is subjective, it is important to have these descriptions to streamline the discussion and provide a limited number of priorities to initially focus on.

Once four to eight opportunities (ideally, one or two tactical implementations for each foundational element) have been prioritized, the team should assign leaders for each opportunity, who will work with the organization to implement them in a pilot capacity. To keep things at the tactical level, there should be limited levels of approval for the pilot and a quick turnaround on evaluating and communicating the results. My recommendation is to accomplish these in three to four months to be able to quickly move from pilots to a full rollout.

A critical aspect of the tactical opportunities is to communicate to the organization how they are doing. This can be formal, through newsletters and emails, or informal, through one-on-one conversations or other communication channels. It is essential to capture the story behind the pilot and the value created, using the core values and foundational elements to reinforce understanding of and messaging for Quality First. In addition, any failure needs to be included—being as transparent as possible is essential to building trust on your Quality First journey.

Then the Strategic Side

Once you have successfully accomplished your tactical implementations, you are now ready to address the strategic side, as you have built up credibility on the value of continuing your Quality First journey and also have a better idea of what to focus on. The strategic areas that typically need to be addressed include:

- Core values: As you went through your core values, there are times when you identified the need to change/update the communication of your core values to match what they actually are. My recommendation is that you start with the tactical implementations (if not accomplished during the pilot phase) for the updated core values to better understand current and future states and then use the results of the tactical implementation to communicate the change in core values. The strategic initiative is then about implementing the communication, gaining alignment with the new core values, and getting those values used in the everyday vernacular.

- Data: Many organizations have an overwhelming amount of data but very little data strategy. The focus of this strategic initiative would be on the understanding of your data (that is, data map), the identification of key metrics, and likely the formation and use of a data lake to drive understanding and changes. Information on identifying key metrics and changes to those metrics from the tactical implementations can streamline this strategic initiative and provide the value proposition to move toward a data lake approach.

- Knowledge: Even fewer organizations have a knowledge strategy. This strategic initiative mirrors that of the data one, in that you need to create your knowledge map, understand the current creation and use of knowledge, and analyze how it is embedded in your processes and systems. This should likely lead to a knowledge lake approach.

- Individual: While most organizations say their most valuable resource is their people, not all of them behave that way. Therefore, this strategic initiative is focused on the individuals, what their value is, how they are recognized and rewarded, and how the organization drives value and continuous improvement through the individuals by defining and doing RT1.

Even though these are called strategic initiatives, they may or may not rise to the level of the traditional five-year strategic initiatives that most companies plan and implement to improve their organization. My preference is that they are not such elaborate initiatives, so you can address them sooner and hopefully take fewer years to accomplish the change. Ideally, these should be one-year strategic initiatives that have sufficient leadership and resources to accomplish RT1 in that time frame.

A 1–5–10 Plan

At this point in your Quality First road map, you should be able to create your 1–5–10 plan, which establishes your road map and milestones for the first year, the fifth year, and the tenth year. As with most concepts in the Quality First journey, this plan should be simple and manageable. I personally like to go with graphical representations with some description—typically, one page for the graphic and one or two pages for the description for each of the three milestone years.

The First Year

The first-year plan assumes that you have already accomplished the core value and foundational elements workshop and are through the initial tactical implementations to gain a clear understanding of where your organization is and where you need to go. Therefore, the key goals for the first year are to get your organization educated and aligned behind Quality First and to get through the initial strategic initiatives:

- Communication—As with any major change initiative, communication is key to getting your organization to understand the change, get behind the change, and then actually change. Therefore, it is essential that you create a Quality First communication plan that will keep your Quality First journey at the forefront of the company throughout the first year. Key elements of the communication plan should include:
  - Kick-off: Clear, concise, and consistent messaging during the kick-off of the change is needed not just from the CEO but also from all the core senior leaders. There needs to be alignment on the message and where you are going. In addition, this cannot be a "single announcement;" it needs to reach each individual in the organization at least seven times for it to be heard and understood.

- Periodic: At least on a monthly basis, additional communications should come out about the progress being made. This must include what is working, what is not working, and what is being changed to overcome what is not working. Being open and transparent will bring people in more quickly to and through the changes.
- Visual: The more visual and graphical you can make the messaging, the more likely it will be received properly. This could simply be where you are on a graphical road map or posters highlighting the key messaging.

- Strategic initiatives—The primary focus during the first year is to implement the key strategic initiatives required to transform your organization to become Quality First. As introduced, these initiatives should address changes or improvements in your core values, data, knowledge, and individuals. A great way to approach and message these strategic initiatives is that they are the catalysts required to energize and accelerate the transformation to Quality First.
- Vernacular—The third focus area for the first year is to get the Quality First vernacular to be used throughout your organization. This is primarily focused on the foundational elements and can be easily accomplished by having leadership include the vernacular in meetings and one-on-one interactions.

Throughout the first year, the Quality First leadership team should meet to evaluate progress and adjust as you implement the initiative. I recommend meeting monthly, but the frequency will depend on your organization and the specific plan.

The Fifth Year

After the first year, you will continue to focus on your Quality First journey but will begin to transition to reinforcing the use and integration of the foundational elements throughout the organization and consistently communicating the everyone benefits performance element. Therefore, for the two- to five-year plan, your efforts should be organized around each of the foundational elements. There will be variability based on how your first-year rollout went and the progress that was made, and the transition may be at 18 months instead of the one-year mark, but it should not be much later. You will have covered:

- RT1—Hopefully, by this point, you have been defining and doing RT1 throughout your organization and have identified and communicated successes. During this phase of your transition, your focus should be on identifying opportunities to use RT1 to improve a project, department, or the organization and to continue to capture and share successes. In addition, it is essential to identify and address any failures in the organization when RT1 was not achieved—this is not to be taken as a negative but as a learning opportunity for the individuals and the organization.

- Individual/Collaboration—A primary focus during this phase will be to work with human resources, information technology, and the leadership on continuing to drive focus on the individual (only they can achieve RT1) and how individuals collaborate to improve as a whole. This will include structural changes to how you identify, onboard, train, and even off-board your employees. In addition, significant efforts should be made to achieve knowledge at the point of need so each individual not only understands what RT1 is for his/her work but also has the tools and knowledge to achieve RT1.

- Embedded Verification—Similar to knowledge, significant efforts will be required to embed verification into each work process so it is just part of what individuals do and is not a burden to their work. This is likely the hardest change to make, as many organizations will get too detailed. The best approach is simplicity, which is in many cases simply asking the individual at key milestones if he/she achieved RT1. Not getting into the details provides flexibility as the definition of RT1 changes. Regardless of how you embed verification, the key to long-term success is having someone evaluate how the verification is being accomplished and analyzing the resultant data for trends and areas to focus on (issues).

- Continuous Improvement—the key to continuous improvement is to focus on getting everyone to improve what they do every day. This does not discredit the need for larger process improvements, but daily, continuous improvement is where you'll get the greatest gains. This daily improvement needs to be a clear expectation from leadership and have ongoing reinforcement through messaging, success stories, and leaders demonstrating their daily improvements.

- Everyone Benefits—At least annually, the Quality First leadership team needs to accomplish a benefit review to identify and document how everyone is benefiting by transitioning to a Quality First organization. While likely highly subjective, just having this discussion and sharing the results with the organization provides clarity on why you are moving toward Quality First and the value it brings beyond the organization. It also helps the leadership team identify issues that are hampering achieving an everyone benefits result.

At the end of the fifth year, your organization should have fully integrated the foundational elements into the way it plans and operates all aspects of its operations.

Everyone in the organization should recognize that Quality First is coming from its core, is providing value to how they do their work, and is fully integrated into how everything is accomplished throughout the organization.

The Tenth Year

As you move past year five and Quality First is part of your culture, you enter what I consider the hardest phase of Quality First. Think about it: You have integrated the foundational elements, you are continually improving the definition and doing of RT1 by the individual, and you have removed many past issues and wasted efforts from your systems. At this point, things are "just working"—which is the challenge.

When things work, the natural tendency is to forget why they are working, and especially new people will begin asking why you do things the way you do them. It is at this point that you could start sliding back to old ways (or new ways, to some) and Quality First begins to fade.

To overcome the natural tendency to forget and slide backward, throughout years six to 10, the focus for Quality First is on reinforcement and continuing to communicate value to the organization:

- Quality First reinforcement—A primary driver of an inadvertent change in culture is the addition of new people who don't align with your culture, which is especially challenging when talking about quality in a broader context, due to so many different perspectives and experiences with quality programs. Therefore, to maintain understanding and commitment to Quality First, there are several opportunities to reinforce its structure and use:
  - New employee onboarding: During the hiring process and especially when onboarding new employees, you must educate them on your core values and the foundational elements. Specific attention needs to be given to defining and doing right for each employee's scope of work.
  - Annual celebration: Organizations celebrate a lot of things, but they don't often celebrate quality processes. To maintain momentum over the years, there should be an annual celebration that recognizes the improvements that have been made; reinforces the foundational elements; demonstrates the use, creation, and sharing of data and knowledge; and discusses failures and the lessons learned from them. This celebration can be at the organizational, department, or team level, or a combination of all three.
- Communicating value: In concert with the annual celebration, it is important to integrate Quality First into your organization's communications, highlighting successes throughout the year, and creating annual

summaries to share both internally and externally. External communications are important at this phase because they help differentiate your brand from others through transparency and openness.

Core Tools to Deploy

Throughout this book, I have reviewed several tools you can use to improve the definition and implementation of Quality First in your organization. It is important to understand that these are tools to help you and that they are complementary, but not essential, for becoming Quality First—so use which ones make sense to you and your organization.

Agile Teams/Collaborative Planning

The biggest change I have made over the past decade has been the use of agile teams and their use of collaborative planning. Those doing the work plan their work, so the team members hold themselves accountable for getting the work accomplished.

The basis for the concepts of applying agile teams and collaborative planning come from Jeff Sutherland's (2014) *Scrum: The Art of Doing Twice the Work in Half the Time* book, where he provides great background and detail on the value of scrum (the use of agile teams and elements of collaborative planning) and the challenges and issues with waterfall scheduling (a traditional planning approach).

If those doing the work are the ones planning the work, how do they do that? This is collaborative planning, which has many names and iterations out in the market (pull planning being the most recognized). My take on it is simple:

- Room setup—Start with a large, open wall that team members can gather around and put sticky notes (activities) on.
- Start with the milestones—Wherever fixed in time, start with the milestones you are working toward. This could be long-term milestones (years out) or short-term milestones (weeks or months out).
- Individual activities—Have each participant write down individual activities that need to be accomplished to meet a milestone. These should be placed in rough chronological order.
- Group review—Have the group review the individual activities, reorder them in sequence, and combine or add activities as needed.
- Streamline activities—Have the group start at the milestone and work back to the beginning to identify opportunities to streamline the workflow—reduce activity requirement, duration, or concurrency with others.

The most valuable result of collaborative planning is that the individuals doing the work get a greater understanding of the activities that occur before, after, and concurrent with their activity. When the work actually gets accomplished, they are more likely to reach out to others at hand-off points and when issues arise compared with other planning methods.

While collaborative planning establishes the general activities and flow of work, agile teams accomplish the work as effectively as possible. There are three important elements for agile teams:

- You can only do one thing—While we have all tried it and continue to try it, multitasking just does not work. You are less efficient, the deliverables are not as good, and you have much higher stress when multitasking. Therefore, in agile teams, each individual focuses on doing just one activity at a time. This is accomplished by using a "to do–doing–done scrum board" (physical or virtual) that contains all the activities. Those activities that still need to be accomplished are placed in the "to do" column, which, at the start, includes all activities. When an individual is working on an activity, he/she moves it to the "doing" column. This indicates to everyone that the activity is being worked on and who is working on it. An activity can move from "doing" back to "to do" if it was not completed and the individual had to move on to another activity, or it can be moved to "done" when it is complete.

- Sprints—The classic approach to agile teams is that they work on sprints—typically, two to three weeks in duration. They start the sprint by putting all the activities to be accomplished by the team in the "to do" column. Key to the sprint's success is that the team does not allow new items to be added, as that shifts them away from achieving their goals and timelines. While things do change, unless the overall scope has drastically changed, this approach of not adding activities is very effective and should be maintained.

- It is a team effort—The third element is that the sprint is a team effort and that the activities are assigned to the team, not individuals. While individuals will ultimately do the work for the activity, the team is responsible. In agile teams, this enables any individual who can do the work to do the work. For example, if team member A finishes an activity early, he/she can look at the "to do" list and choose what to work on next. Also, if team member B is out sick, any of the other team members can jump in and take over an activity.

A key subtool of agile teams and collaborative planning is the consistent use (daily) of stand-up meetings, which were discussed in detail in Chapter 4, Collaboration Tools. These should be 10 to 15 minutes in duration and quickly cover what worked/did not work from yesterday and what the focus of today is and what can be improved today.

There is obviously a lot more detail about understanding and implementing agile teams and collaborative planning. However, if you follow the basic guidance given above, you will experience most of the benefits these can bring to your teams and your organization.

One-Page Communication

In our current digital age, the art of communication has been lost. Either the information is not enough (a short text message) or it is way too much (a 100-page action plan). The problem with either end of the spectrum is that the information is not likely understood, and, of more importance, the recipient of the information does not know what action needs to be taken.

A great tool that I recommend teams adopt is that of one-page communication, which is based on an informational A3 approach; however, it does not have the rigidity of traditional A3 problem-solving techniques. The key elements of a one-page communication are:

- One page—It sounds obvious, but all you want and need to communicate is on one page. While you can use larger-size pages to convey more information, I find that the largest page size you need is 11 inches x 17 inches (A3 in metric).
- Not just text—As you have limited space to convey your message, it is often best to include graphics, charts, and simple tables to convey key information. This also provides better understanding and retention than just text on a page.
- Clarity of purpose and action—Through titles and sections, just by looking at the one-page communication, recipients should be able to understand its purpose as well as the action they need to take after receiving it.

Creating effective one-page communications is not easy—it's not just cramming a bunch of information onto one page. It takes time, often several iterations, and maybe even a few cycles of review to finalize. However, by having teams go through the process of creating a one-page communication, they not only understand the issue or update better, but they are also able to talk about it more effectively because they have had to get to the core of it all.

As you roll out one-page communications in your organization, be careful that you don't go overboard. Not everything needs to be or can be consolidated down to one page. A 100-page report is still a 100-page report. However, a one-page communication conveying actions to be taken that complement the 100-page report may be appropriate. In addition, the organization's leaders need to set an example of effective communication. If they are asking for something from their team, can it be delivered in a one-page format?

Nominal Group Technique

As mentioned earlier and used multiple times, I love the NGT as a way to brainstorm and reach a consensus on just about anything. I have used NGT in any situation to identify and prioritize ideas, to establish organization or project goals/guiding principles, or to decide which direction to go. It is a very effective and flexible tool.

As was introduced in Chapter 3, on the right foundational element, the NGT is composed of four steps:

1. Brainstorming—Each participant writes down his/her responses to a question on a piece of paper without any discussion. The NGT facilitator then goes around the room (physical/virtual) and records a response from each individual in a round-robin fashion (one at a time), without any discussion.

2. Discussing—After all responses are recorded, the facilitator goes through each response to clarify what it means and ensure the entire group understands and agrees with it. While items can be combined, they cannot be deleted. Items can also be added as people discuss responses.

3. Voting—Each individual gets five votes, where he/she identifies and ranks his/her top five items.

4. Summarizing—The facilitator consolidates the votes and reviews the results with the participants, typically focusing on the top 10 items to get concurrence that it represents the key items.

As you implement the NGT in your organization, there are a few tips and tricks to help you become more effective and efficient in its use:

- Diversity in attendees—The ideal size for an NGT workshop is no more than 20 people. To get the most out of the workshop, try to maximize the diversity of levels and viewpoints of the participants. The broader the perspectives, the more likely you will get to the best result. Consider bringing participants from outside the organization if relevant.

- Two facilitators—It is best to have two facilitators. There is the participant-facing lead facilitator who manages the process, records their responses, and drives the discussion. This person is at the front of the room. The second is the documenting facilitator, at the back of the room, who not only records the responses but also summarizes the discussions. The value of the documenting facilitator is that at the end of the workshop, you have the items prioritized along with a narrative summary of each response, making it much easier to determine the actions to be taken.

- Paper flipcharts—It is a bit old school, but paper flipcharts work best to record the responses and then add the ranks to them. It is the quickest way to allow everyone to see the 30 to 50 responses for voting.
- One at a time—The lead facilitator needs to be strict about only having one person talk at a time and not having discussion as responses are recorded. This prevents quiet individuals from being shut down by more aggressive individuals. The facilitator's role is to get input from everyone.

Data Intelligence

I covered data intelligence in Chapter 9, but it is worth summarizing here. Connecting, analyzing, and understanding data from across your systems is the untapped resource in most organizations. Maximizing the value of this resource will be a key differentiator for organizations in the future; those that drive with data will be far ahead of those that do not.

As a reminder, to find the data and metrics that will help you understand your organization and drive Quality First, you need to be inquisitive, patient, and resilient—you will not likely find them on your first try. Start your investigation where you are hurting, where your processes are not working, or where things appear to be working, but no one can tell you why. In these areas, you will find the data that show status and trends where you need to focus your efforts.

One last thing on data and issues: sometimes, fewer issues are not an indication of better, especially when issues are not being reported. This can be problematic when the issues being reported are not mandated by a strict policy or law. Therefore, a key element of data intelligence is not just the data map but also how consistent and reliable the data are on the map.

Quality First Leadership

For Quality First to succeed and thrive in your organization, there needs to be leadership at several levels—it cannot strictly be a top-down or a bottom-up approach, or even a single leader driving the change. Quality First is coming from, but also often crystalizing, your culture. Without getting leaders at all levels engaged and driving, Quality First will not stick and withstand future degradations to its implementation and value. There are three key levels of Quality First leaders: C-suite/board, department heads, and champions.

C-Suite/Board

The most senior leaders in the organization, the C-suite and those on the board, must be fully engaged and aligned with transitioning to become a Quality First organization. The challenge I have found with any major shift in how a company operates is that while the leaders will say they are committed, they do not actively demonstrate their commitment or change the way they operate personally to achieve the new operating state for the organization. At best, they do nothing; at worst, they sabotage the change.

Therefore, it is essential for the success of Quality First that senior leadership be onboarded to Quality First and engaged throughout the transition. This can be accomplished in many ways, and it often depends on the culture of the organization which way works best. However, your approach to onboarding and engagement should include these key activities:

- Make Quality First personal
- Senior leader core value/right workshop
- Current/future state heat map
- "I commit" exercise
- Senior leadership operational changes
- 1–5–10 plan engagement

Make Quality First Personal

The onboarding of senior leadership is one of the biggest challenges for an organization starting its Quality First journey. This is primarily because these leaders are extremely busy, often seeing themselves as not having time to change; in many cases, they feel they are already doing great. Therefore, I find that onboarding your senior leaders requires more of an experiential approach than an educational approach—they don't need to understand Quality First, they need to feel it.

To start the process, there needs to be some level of understanding of the Quality First framework and approach. For all leadership teams, and especially small ones, I recommend simply doing a group read and discussion of this book. The key is that over two to three months, the leadership consistently meets (physically or virtually) to discuss one or two chapters and how it applies to their organization. However, when a book club is not possible, providing a one-page primer is a great start (see Figure 11.3).

QUALITY FIRST PRIMER

What is Quality First?
Quality First supercharges your organization from your core values through integration and consistent use of the Foundational Elements

FOCUS ON 4 - 1 - 3

Leading Your Organization to be Quality First requires focusing on the 4 - 1 - 3 components.

4 Fully integrate the 4 Foundational Elements into your organization starting from and through your core values:

1. Right
2. Individual/Collaboration
3. Imbedded Verification
4. Continuous Improvement

1 Highlight and communicate the Performance Element of Everyone Benefits to demonstrate and drive success.

3 Utilize the 3 Catalysts to accelerate your Quality First Journey:

1. Power of Once
2. Data Intelligence
3. Knowledge Snippets

ROAD MAP

Create a simple road map to take you on your Quality First Journey, focusing on 1 - 5 - 10 year milestones.

KEY TOOLS

- Agile Teams and Collaborative Planning
- 1-Page Communication
- Nominal Group Technique
- Business Intelligence

LEADERSHIP

Fully engage leadership throughout the organization, from C-Suite/Board to Department Heads and Champions.

Figure 11.3 Summing up Quality First.

Once the leadership team understands Quality First, you then need to make it personal for them. I find that a simple NGT workshop provides a quick way to get the senior leaders engaged in a discussion, as well as having them prioritize the change and improvement they can make to the organization. The workshop should be composed of three questions, and you can take a pro or con approach to the workshop:

- Pro approach—look at positives
  - How do we do RT1?
  - How do we support our individuals to do RT1 every time?
  - What are examples of continuous improvement?
- Con approach—understand current challenges
  - What are examples of not doing RT1?
  - How do we not support our individuals to do RT1 every time?
  - What are examples showing where we could improve how we operate?

You can also blend the two, such as "How do we improve doing RT1?" Regardless of your approach, for the voting and ranking of each question, you are asking leadership to vote on what they want to continue to do (pro) or what they need to change the most (con). At the completion of the workshop, you have alignment from senior leadership on how they define the concept of RT1, support of individuals, and continuous improvement. (You could also do embedded verification, but I find this is best handled at levels of the department head and down.)

The primary value of using the NGT workshop approach is that every senior leader participates (remember—you get their input without discussion), and they get to vote on what is important to them, with everyone having the same number and value of votes. It is very important at the end of the workshop to have everyone verbally state that they agree with the overall priority and not their personal vote—this avoids, or at least minimizes, sabotaging of focus after the workshop.

Senior Leader Core Value/RT1 Workshop

Once senior leadership has a more personal alignment with Quality First, the next step in onboarding them and the organization to Quality First is to have the senior leaders participate in your core value/RT1 workshop, as covered in Chapter 3, where you establish how RT1 comes from and drives your core values.

It is essential that senior leadership be involved and actively participate, as you are clarifying and crystalizing your core values (remember—your core values already exist, and they are how you operate, but they may not have been clearly communicated). It will be their involvement in documenting your core values and then the definition of RT1 that will create a solid foundation from which to drive Quality First throughout the organization for years to come.

Current/Future State Heat Map

I find that while everyone likes to know where they are and where they are going, senior leadership is especially attuned to needing a current- and future-state understanding. As I love visuals, using an X–Y graph to show where you are and where you want to go is a great way to communicate the journey. This Quality First heat map is a simple tool to use to indicate the current state of the key opportunities (or issues) and where you can take them in the future (two to five years)—see Figure 11.4.

While often very subjective, the Quality First heat map is important, and it simplifies what you are going to focus on and visually shows that there is great room for improvement from the current state. It is the delta between the two that will get leadership and the organization excited about Quality First and where it can take you.

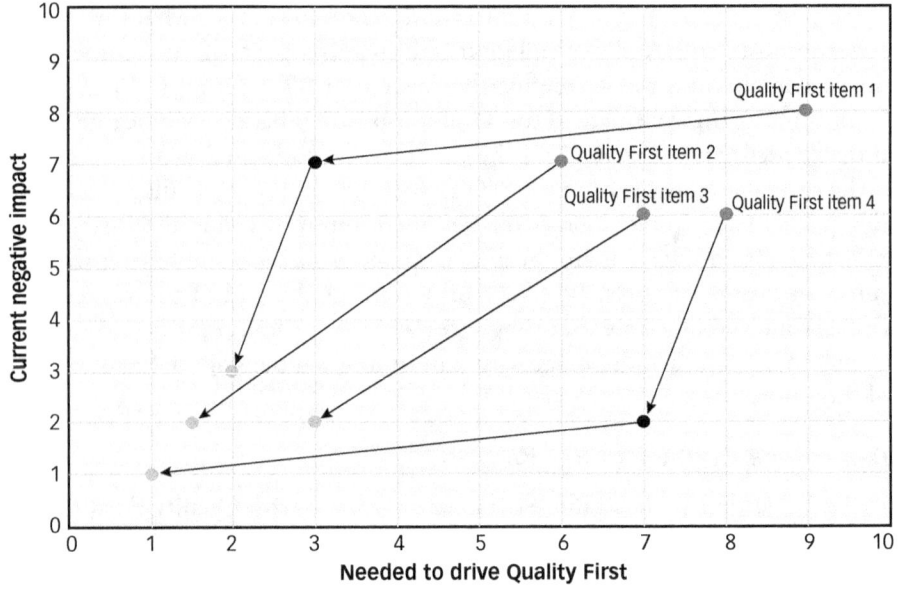

Figure 11.4 Quality First heat map.

As you see in Figure 11.4, the horizontal axis represents the "need to drive Quality First," with a high value indicating you need to address the item in order to become Quality First. The vertical axis represents the "current negative impact," with a high value indicating a large negative current impact on the organization. Therefore, the goal of addressing each of the items will integrate Quality First into the organization and eliminate the current negative impact on the organization—moving down and to the left, as you advance Quality First.

However, the journey may not be a straight line. For example, the Quality First item 1 in Figure 11.4 is required and is having a large negative impact on the organization, yet the immediate focus is to get the concept implemented, which has some, but limited, effect on reducing the negative impact. This is typically true for the RT1 foundational element, in that you need to first get the concept integrated into your organization before the negative types of impact are eliminated, by defining and doing RT1.

In the case of Quality First item 4, you can reduce the negative impact very quickly, yet there is still a strong need to drive the item to integration to become Quality First. A great example of this is the continuous improvement foundational element, where the benefits are quickly received by doing daily stand-ups and focusing on improving everyone, every day. However, it takes a long time to integrate this into your operations to the point where it is just second nature to do so by everyone in the organization.

"I Commit" Exercise

The next step for senior leadership is to advance the idea of making it personal, with each leader going through an "I commit" exercise. This is simply having each leader write down what he/she personally commits to accomplish over the next 12 months with respect to Quality First—using all the information created to this point. The format I find works the best is:

- I commit to [doing something]
- by [time frame]
- in order to [what value provided]

Ideally, this is done with and by the entire senior leadership group, where each leader shares and discusses his/her commitment. This is meant to enable support and accountability of everyone to their commitments and not as a way to punish or degrade a leader who does not achieve his/her commitment. Think of this from the agile/scrum board perspective—each leader has one item he/she is either to do or is doing. If, for some reason, the leader cannot get to the commitment, the other leaders can and should jump in to provide support to accomplish the item.

Senior Leadership Operational Changes

This then leads to the next step for senior leaders—they themselves need to change how they operate, adopting the Quality First foundational elements and the key tools introduced. The main reason for this is to think about the shadow of leaders.

A small action by a leader ripples across the organization as his/her shadow spans it. If the leaders change how they operate, it sends a clear and very loud message to the rest of the organization that they are taking Quality First seriously as well as leading the way in changing themselves.

Put another way, Quality First affects the entire organization and is a do-as-I-do approach, not a do-as-I-say approach. The key changes to senior leadership's operations should include:

- Vernacular—It is essential that senior leaders understand and use the Quality First vernacular in their meetings and interactions with others. Primarily, this includes the foundational elements. Simply asking how an individual understands and does RT1 will go a long way toward others focusing and implementing the foundational elements in everyday activities.
- Senior leadership agile team—Having the senior leadership adopt a "to do–doing–done" scrum board provides immense value to the organization. First, the senior leadership team will begin to operate more as an agile team instead of as individual senior leaders, better supporting and aligning with the needs of each other. Second, there is clarity on priorities and what the senior leadership team needs to do above and beyond their specific roles.
- One-page communications—Look at what the senior leadership team is asking from others and identify which ones could be streamlined to use the one-page communication approach. These could be monthly or quarterly updates, business-planning efforts, or even succession planning. Moving to the one-page communication approach for key activities sends a clear message to the organization that you expect better, more concise, and more-thought-out communication with senior leadership, as well as with everyone else. Also, by switching, you are reinforcing the expectation of continuous improvement—not just repeating what was done the past 10 years on reporting.
- Data intelligence and knowledge snippets catalysts—Each senior leader in the organization needs to understand and personally drive the data intelligence and knowledge snippets that are catalysts of Quality First. At a minimum, this should be a standard agenda item, should have identified leaders, and should be part of the senior leader scrum board of ongoing to-do activities.

For some organizations, these changes will be easy; for others, they will mean substantial change. What is important is that you commit and align as senior leaders to change how you operate and continue to improve how you operate. Use your long shadow across the organization as an example for others to change.

1–5–10 Plan Engagement

The last part of senior leadership is their ongoing engagement in your Quality First transformation. They need to be engaged throughout the transition as well as taking leadership roles for doing the transition. Therefore, a key part of the 1–5–10 plan is how and when senior leadership will be engaged in the progress and transition. These key items should be included:

- Live status reports—A goal must be that updates or status reports on the Quality First transition need to be live—they are just there—and they do not need to be specially created for leadership every once in a while. Yes, this is hard to do, but by focusing on the data intelligence catalyst (Chapter 9), you can identify the metrics that "come from your systems" that can be easily, and automatically, shown as a one-page communication (dashboard) to which senior leadership will always have access.

- Standard meeting agenda item—Quality First needs to be part of every senior leadership group meeting and should be one of the first items discussed—and definitely not the last. Having a discussion early in every meeting gives Quality First the priority it needs, and an early focus on action items helps avoid being overcommitted by the end of the meeting to effect the needed change.

- Annual review—Each year, the senior leadership group needs to take a deep dive into Quality First, update the 1–5–10 plan, and accomplish a new "I commit" exercise. This re-establishes the baselines for where you are and where you are going.

Department Heads

In small organizations, department heads are often part of the senior leadership group. However, as the organization grows, the senior leadership team adds more regional leaders, and typically not all department heads remain part of the senior leadership team. In addition, even when they are part of the senior leadership team, they have additional focus items as department heads. This also applies to any senior leader, for they need to focus on how they operate and interact with the individuals within their department/group.

When looking at and discussing "departments," this really applies to any group within an organization and what they can do to drive Quality First within the group. For our purposes here, I discuss them from the typical department focus, such as such as human resources, information technology, marketing, and operations.

The approach and rollout of Quality First in a department are primarily focused on the foundational and performance elements and how they specifically apply to the department. The department heads are responsible for the overall rollout of Quality First, but they need to engage and depend on the leaders in the department to actually go through the transition with them:

- RT1—As covered in Chapter 3, each department needs to define and do RT1 for their department. The definition of RT1 should be complementary to the organization's RT1s, but, as demonstrated earlier in the book, the department RT1 statements will focus on the department and how they achieve RT1 for their role within the organization. Therefore, the department head needs to work with his/her team to define the RT1 statements and get the entire department aligned with the statements.

- Individual/Collaboration—Understanding that only the individual can do RT1 is easy. The hard part is fully implementing this in a department where each individual understands what RT1 is and is able to do RT1 for his/her work. I would start with daily stand-ups for each team within the department to engage individuals and enhance collaboration. To drive the importance of the stand-ups and defining and doing RT1, the department head needs to actively participate in the stand-ups on a periodic basis (and the subordinate leaders need to participate in their team's stand-ups, and so forth). If a leader has five teams, he/she could participate in one per day or one per week—it will depend on the ease of participating and the need to reinforce the importance—more frequently at rollout and less as time passes. However, the leader needs to always have some participation; it cannot be zero. The added benefit of this is that the leader has a much better understanding of the status of the department and interacts with everyone within the department, not just through his/her subordinates.

- Embedded Verification—Embedded verification will likely be one of the harder foundational elements to integrate into the department. An easy way to start is to just ask your team members how they can verify RT1 as they do their work, so it's not an extra item but just part of their work. Once you have their input, then you can identify and implement approaches to integrate into workflows and systems. Because most embedded verifications are system-based (this is why it can be so difficult), many systems just don't have verification embedded in them or are not part of the process but are an additional action to take.

- Continuous Improvement—By implementing daily stand-ups in your teams, you will also be implementing continuous improvement. From the department perspective, what is important is to collect and share the stories of improvement. The more the department leadership recognizes improvement, the more individuals are willing to try something new and improve themselves. It is also important for department leadership to approach failures from a learning perspective and to "fail forward."
- Everyone Benefits—As mentioned many times, everyone benefits is a result of the foundational elements, but it does need to be nurtured for the benefits to be recognized. At a minimum, there must be an annual discussion of how everyone has benefited through the foundational elements and Quality First.

Champions

The last layer of leading the Quality First transition will be your champions—those individuals who rise to the top as passionate leaders of change and improvement. As these are champions, anything related to Quality First will likely be above and beyond their normal duties, and it is important to recognize and support their efforts. To support, encourage, and recognize your champions, consider these points:

- Champion team—You need to have a means for the champions to come together and discuss what they are doing, what challenges there are, and how to share best practices on Quality First. The team should meet at least quarterly but be in constant contact through a group site/system.
- Leadership—The team should self-lead (think agile) but needs to have a senior leader sponsor. My recommendation is the CEO, for this provides immediate access to the organization's leader but also provides instant credibility and recognition of the champion team.
- Recognition—Consider ways to recognize champions and their contributions to your Quality First transition. This could be as simple as highlighting in a communication or meeting, to rewarding champions with a trip with senior leadership to combine working on advancing Quality First with enjoying downtime/facetime with leadership.

It is important to recognize the majority of change, and the overall Quality First transition, will happen with and through your champions. This is simply the fact that change occurs at and with the individual. Tapping into and supporting your champions is your fastest way to Quality First.

Conclusion

Congratulations! You have made it through *The Leader's Playbook*. I hope you enjoyed reading the book and that you are ready to start your Quality First journey—if you have not already started. In closing, here are a few thoughts I would like to leave with you as you embark on your journey.

Quality First Must Come from Your Core—Not a Layer or Swim Lane

Throughout, I have emphasized that Quality First must come from and drive your core values. By coming from your core, quality transcends all aspects of the organization and becomes who you are and what you do. Put another way, quality is not an add-on activity, and it is not a layer or swim lane within the organization.

This can be a hard concept, and it can be complicated to achieve in highly regulated industries or industries that have large and strong quality departments. To be clear, I am not saying to get rid of quality departments! However, if you look at the foundational elements, most quality departments do not bring quality to an organization; they help document its quality. This is especially true in highly regulated industries, where thousands of inspections are accomplished to document the achievement of "quality" yet rarely define what RT1 is or enable the individual to do RT1 and document it through embedded verification.

What you will find when you become Quality First is that your quality department becomes much more effective in their roles. They are no longer the police finding mistakes but rather a trusted partner in defining and doing RT1 by the individual so they can continuously improve.

Understand and Celebrate Value

To keep Quality First coming from your core and advancing requires recognizing the progress being made through Quality First. By implementing the foundational elements and using the data intelligence and knowledge snippets catalysts, you will come to understand the value you are generating through Quality First. The next thing you must do is celebrate the *value*.

However, it is important to understand what you currently celebrate, which may not be obvious or marked with a typical celebration such as a party. It could be how you determine raises and promotions, or where you invest your capital in the organization. Understanding how you celebrate will let you recognize how your current celebrations do or do not align with Quality First. A few ways to recognize and celebrate Quality First are:

- Best Quality First advancement: Have a competition to find the best advancement of Quality First in your organization, which could be around a single or all the foundational elements. This could also recognize individuals or teams.

- Bonuses: While money is not everything, it does get most people's attention. Tying and stating Quality First advancements to bonuses is a great way to see the value of becoming Quality First. As you progress and get better data and knowledge, you will be able to say that 50 percent of the organization's improvement in profit was due to Quality First (RT1, once, and continuous improvement).

- Personal performance plans: Making the foundational elements part of an individual's personal performance plan allows managers to recognize individual improvement throughout the year. Having goals and means to reach them is key to any improvement initiative.

- Quality First events: Your organization and departments already have events for meeting sales targets, customer satisfaction, and the like. Adding Quality First to these events raises awareness of its importance to the company and also can supplant one or more other events you currently host.

Share Your Journey with Others

My final thought and guidance for you is that it is important to share your journey with others. This is why I wrote this book—to share what I have learned that has worked and improved the lives of thousands of people. Now it is your turn. As you embark on your Quality First journey, share with others. Invite them to come to your organization and see what you are doing.

This isn't about just showing the good stuff. You must also show what isn't working or has not worked, and what you learned from those situations. I know I learn more from my failures than my successes. By bringing others in, you will find that it gives you passion to improve what you are doing and to learn from others about what they see. This is a great way to identify your blind spots and address them before they severely affect your organization.

The more open and honest you can be with others, the more open and honest you will be with yourself.

Appendices
Quality First Examples

Appendix A
Construction Quality First Example

From 2005 to 2020, the construction industry went through a quality renaissance, starting with five organizations truly integrating quality from their core and going on a Quality First journey, with dozens more following. The result has been a 40 percent improvement in their overall ranking in the industry's top 400 ranking compared with others, as well as two times the growth in revenue.

In looking at the construction organizations from a Quality First perspective, there are several commonalities and lessons across these aspects (see Figure A1):

- Trigger
- Foundational elements
- 1–5–10 plan
- Catalysts
- Sharing with others
- Achieving Quality First

The Trigger

For the majority of construction companies, there was a trigger in the early 2000s that led them to move toward Quality First, which were large financial and image losses due to major rework and lawsuits on completed projects. While the original impetus and goal of the companies was not Quality First, in retrospect, that is what happened due to the outside leadership brought into the various companies to define and drive quality in the organization.

Figure A.1 content

Trigger
Major losses from rework and lawsuits

Foundational elements

RT1
Do each scope of work once to meet contract documents

Individual/collaboration
Focus on individual doing design (contract documents) and installing the work (trades)

Embedded verification
Individual doing work documents achieving RT1 (scope of work) using project systems

Continuous improvement
Feedback loops on wrong to ensure RT1 going forward

1–5–10 Plan

1st year
Focus on individuals doing loss scopes of work:
- Define RT1
- Verify RT1
- Drive learning

5th year
- Expand to all scopes of work
- Share metrics across industry
- Change systems to embed verification

10th year
- Expand to all departments
- Embed foundational elements into procedures
- Quality First value discussion

Catalysts

Data
- Industry-wide proactive and reactive metrics
- Tie systems together
- Leadership engagement in exceptions

Knowledge
- Point of need—each scope of work activity
- Communities of practice
- 1st five-year employees

Sharing with others

Construction quality executives council of 35+ peer design and construction companies
- Annual conference
- White papers

Achieiving Quality First
- 200% greater growth[1]
- 170% greater growth[2]
- 350% greater growth[3]

[1] Compared to Top 400 organizations
[2] Compared to Top 25 organizations
[3] Original five Quality First organizations

Figure A.1 Construction Quality First.

1–5–10 Plan

The approach to the 1–5–10 plan was pretty simple. In the first year, the bleeding had to be stopped—which was to eliminate the losses from the high-risk scopes of work simply by defining and doing RT1 and driving knowledge (best practices) from what was learned.

As the organizations moved past the first year, the five-year plan focused on expanding to all other scopes of work and sharing metrics across the industry to understand where to focus their broader attention. There was also a drive to embed verification in each organization's systems to make it part of the process instead of additional activities using separate systems.

Past the five-year mark, the majority of organizations were taking Quality First to their organization and affecting all key departments. This was accomplished by integrating the foundational elements (slightly different terminology in each organization, but that's OK) into organization procedures and by having Quality First value discussions across the organizations.

Catalysts

Early in their Quality First journeys, the organizations realized the importance of the catalysts' data intelligence and knowledge snippets. For data intelligence, it was about tying data from disparate systems together to better understand the whole and then sharing key metrics across the industry to identify unforeseen risks from a single organization. This actually led to the realization that organizations using their data from a Quality First perspective could identify and mitigate systemic risk from construction projects two to four years ahead of the insurance industry receiving claims from non Quality First organizations. Further, with data, there was a clear shift on reporting data to driving action from the data—this was accomplished by getting leadership to focus on exceptions in the data, which were areas that should not have happened if Quality First had been fully implemented.

For knowledge snippets, there was a concerted effort to drive knowledge to the point of need, focusing on each activity within each scope of work, and communicating what was important to know and do with each activity. There was also the realization that each scope of work would be managed by individuals new to the industry, without knowledge or understanding of the risk. This led to the creation of communities of practice and core process and technical knowledge that new one- to five-year employees needed to know.

Sharing with Others

Starting with a group of five organizations on a Quality First journey, a peer group (Design and Construction Excellence Exchange) was formed to share metrics, best practices, and challenges with each other. This has subsequently grown to more than 50 design and construction organizations. To share beyond the exchange organizations, the exchange publishes white papers, engages with other groups, and convenes an annual conference.

Achieving Quality First

Comparing those on a Quality First journey with the general industry (400 organizations), the results on growth are quite amazing:

- 200 percent greater growth—compared with top 400 organizations
- 170 percent greater growth—compared with top 25 organizations
- 350 percent greater growth—original five Quality First organizations

Appendix B
Manufacturing Quality First Example

There have been hundreds of books written about quality in manufacturing and how to implement a program and even a lot about culture. However, how do we know when a company has quality at its core, or, more importantly, what happens when it does not? To answer this, we will look at an industry's quality ratings and how they change over time for Manufacturer A and Manufacturer B, compared with the industry average—as shown in Figure B.1.

Now, while industry rankings have both pros and cons, they do provide a benchmark to compare organizations and how they change over time. The first thing to note in Figure B.1 is that data before 2005 are not readily available; we do know a few key points as shown, but not the details. However, the trends and relative lines have been verified and are representative of what happened to each organization and the industry.

If we start with Manufacturer B, you will notice that it started out as one of the best but had a hard time with consistency, resulting in large swings in rankings and the resultant actual quality of its products. As you can see, three major events affected its rankings:

1. Late 1990s—The first major shift occurred due to a leadership change at the top that resulted in a significant reduction of funding for its quality department. Before this change, I would have considered the company a Quality First organization, as it had many of its attributes. Unfortunately, this did not come from its core, so it was not sustained through the leadership change.

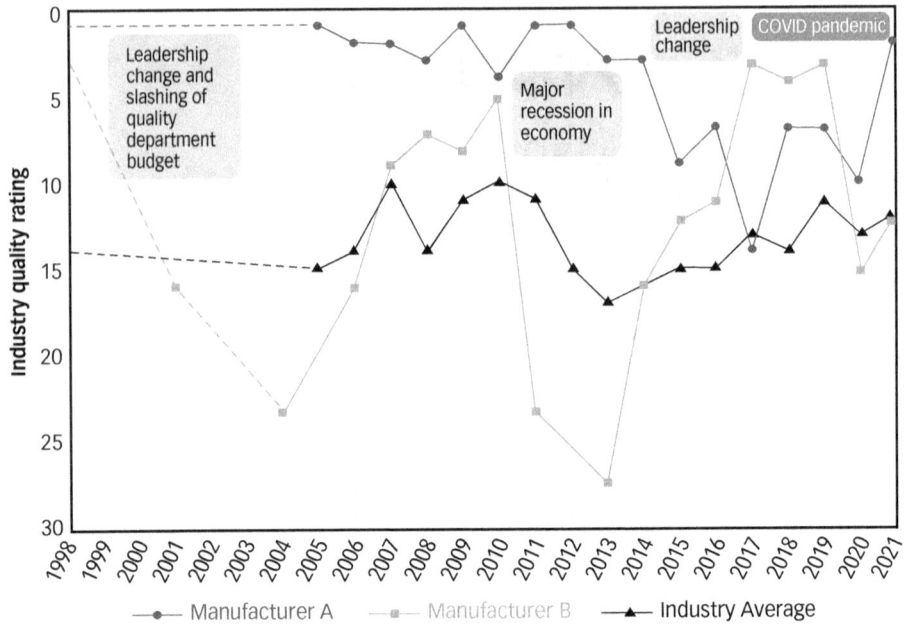

Figure B.1 Changes in quality rankings with major events.

2. Early 2010s—After more than a decade of working its way back to near the top, there was a swift turnaround for the worse when the recession hit the economy and the company. The assumption is that cost cutting affected its quality program and activities the most.

3. Early 2020s—The good news is that in only about six years, it was able to turn back around and get close to the top again. Unfortunately, the COVID-19 pandemic hit, and the firm retreated somewhat but not as badly as before.

Compared with Manufacturer B, Manufacturer A was at or near the top for almost two decades; however, after a leadership change, there was substantial degradation. The difference is that I do consider this a Quality First organization, with quality at and coming from its core. While it did degrade, it did not degrade past the industry average, and it actually improved its performance during the pandemic.

Yes, Manufacturer A did have challenges in going through the leadership change, and that is a word of caution for any of us—during times of stress and other changes, don't let a temporary slide in performance or Quality First derail the entire initiative. Instead, use it as a learning point to enable you to redouble efforts and focus on Quality First. This is the fail forward mentality we discussed earlier in the book.

References

Akers, Paul A. 2011. 2 *Second Lean: How to Grow People and Build a Fun Lean Culture at Work & Home.* Ferndale, WA: FastCap Press LLC.

DCX (Design and Construction Excellence Exchange). 2024. *Measuring Quality in Construction.* https://thedcx.org/measuring-quality.

Delbecq, A. 1975. *Group Techniques for Program Planning: A Guide to Nominal Group and Delphi Processes.* Glenview IL: Scott Foresman.

Hamilton: An American Musical, music and lyrics by Lin-Manuel Miranda, 2015.

Lencioni, Patrick. 2012. *The Advantage: Why Organizational Health Trumps Everything Else in Business.* San Francisco: Jossey-Bass.

Maxwell, John C. 2000. *Failing Forward: Turning Mistakes into Stepping Stones for Success.* Nashville: Thomas Nelson.

Sutherland, Jeff. 2014. *Scrum: The Art of Doing Twice the Work in Half the Time.* New York: Crown Business.

www.ingramcontent.com/pod-product-compliance
Lightning Source LLC
Chambersburg PA
CBHW051400290426
44108CB00015B/2094